Two-Dimensional Electronics and Optoelectronics

Special Issue Editors

Yoke Khin Yap

Zhixian Zhou

MDPI • Basel • Beijing • Wuhan • Barcelona • Belgrade

MDPI

Special Issue Editors

Yoke Khin Yap
Michigan Technological University
USA

Zhixian Zhou
Wayne State University
USA

Editorial Office
MDPI AG
St. Alban-Anlage 66
Basel, Switzerland

This edition is a reprint of the Special Issue published online in the open access journal *Electronics* (ISSN 2079-9292) from 2016–2017 (available at: http://www.mdpi.com/journal/electronics/special_issues/2d_elec_optoelec).

For citation purposes, cite each article independently as indicated on the article page online and as indicated below:

Author 1; Author 2. Article title. *Journal Name* **Year**, *Article number*, page range.

First Edition 2017

ISBN 978-3-03842-492-5 (Pbk)
ISBN 978-3-03842-493-2 (PDF)

Photo courtesy of Mingxiao Ye, Dongyan Zhang and Yoke Khin Yap

Table of Contents

About the Special Issue Editors

Yoke Khin Yap is a professor of physics, and the director of the Applied Physics program at Michigan Technological University (MTU). He earned his Ph.D. in 1999 from Osaka University as a "Monbusho" scholar sponsored by the Japanese government. Professor Yap was a postdoctoral fellow of the Japan Society for the Promotion of Science (JSPS) before his faculty appointment at MTU in January 2002. In 2005, Professor Yap was honored by the U.S. National Science Foundation CAREER Award. Professor Yap has published more than 100 peer reviewed journal articles, and more than 20 book and encyclopedia chapters. He has also edited a book (Springer) and four Material Research Society (MRS) proceedings for a MRS symposium series. In 2011, Professor Yap received the MTU Bhakta Rath Research Award. In 2015, Professor Yap was honored as Global Alumni Fellow of Osaka University.

Zhixian Zhou is an associate professor of physics at Wayne State University (WSU). He received his PhD from Florida State University in 2004. After working at the Oak Ridge National Laboratory as a postdoctoral research associate for nearly three years, he joined the faculty at WSU in August 2007. Professor Zhou's research has centered on investigating the electrical transport properties and device physics of low-dimensional materials, including graphene and two-dimensional (2D) layered semiconductors. His work resulted in multiple highly-cited publications in high-impact journals such as Nano Letters and ACS Nano. Professor Zhou has also been recognized by invitations to present his research at international scientific conferences and workshops as an invited speaker. He is also the recipient of a number of awards including the WSU Career Development Chair Award (2016–2017) and Sultana N. Nahar Prize for Distinction in Research in Physics and Astronomy (2017).

Preface to "Two-Dimensional Electronics and Optoelectronics"

The discovery of monolayer graphene has led to a Nobel Prize in Physics in 2010. This has stimulated research on a wide variety of two-dimensional (2D) layered materials. The coupling of metallic graphene, semiconducting 2D transition metal dichalcogenides (TMDCs) and black phosphorus has attracted tremendous amount of interest in new electronic and optoelectronic applications. Together with other 2D materials such as the wide band gap boron nitride nanosheets (BNNSs), all these 2D materials have led towards an emerging field of van der Waal 2D heterostructures. This book is originally published in Electronics (MDPI) as a special issue of "Two-Dimensional Electronics and Optoelectronics". The book consists of a total of eight papers, including two review articles, covering important topics of 2D materials. These papers represent some of the important topics on 2D materials and devices. Promises and challenges of 2D materials are discussed herein, which provide a great recent guidance for future research and development.

Yoke Khin Yap and Zhixian Zhou

Special Issue Editors

electronics

MDPI

Editorial

Two-Dimensional Electronics and Optoelectronics: Present and Future

Zhixian Zhou [1],* and Yoke Khin Yap [2],*

[1] Department of Physics & Astronomy, Wayne State University, Detroit, MI 48201, USA
[2] Department of Physics, Michigan Technological University, Houghton, MI 49931, USA
* Correspondence: zxzhou@wayne.edu (Z.Z.); ykyap@mtu.edu (Y.K.Y.); Tel.: +1-313-577-2751 (Z.Z.); +1-906-487-2900 (Y.K.Y.)

Received: 18 July 2017; Accepted: 19 July 2017; Published: 22 July 2017

1. Introduction

Since the successful isolation of graphene a little over a decade ago, a wide variety of two-dimensional (2D) layered materials have been studied. They cover a broad spectrum of electronic properties, including metals, semimetals, semiconductors, and insulators. Many of these 2D materials have demonstrated promising potential for electronic and optoelectronic applications.

Graphene has attracted a tremendous amount of attention from the scientific community largely due to its combination of extremely high carrier mobility and thermal conductivity with mechanical strength and flexibility as well as high thermal and chemical stability [1–4]. However, the lack of a fundamental bandgap in graphene has significantly limited its applications in electronics and optoelectronics. The research community is now turning its attention to 2D materials beyond graphene, particularly 2D semiconductors with an appropriate bandgap such as transition metal dichalcogenides (TMDCs) and black phosphorus [5–8]. The interlayer van der Waal bonding in 2D materials also offers the opportunities to create a large number of heterostructures by artificially stacking different 2D materials together without the constraints of atomic commensurability.

2. The Present Issue

This special issue consists of seven papers covering important topics in the field of 2D materials, including two reviews focusing on electronic and optoelectronic devices based on 2D materials. The contents of these papers are introduced here.

In Reference [9], micro-reflectance spectroscopy is used to investigate the differential reflectance spectra of hundreds of MoS_2 flakes grown on a highly-polished sapphire substrate by chemical vapor deposition. This fast and non-destructive characterization technique is able to measure a large number of spectra in different sample locations and in a small amount of time. The growth of smooth and continuous layers of γ-InSe and Sb_2Se_3 layered metal chalcogenide on SiO_2-coated Si and glass substrates by atomic layer deposition (ALD) is described in Reference [10]. This work shows that ALD offers a viable path for producing large area films of metal chalcogenides for future industrial-scale applications. A first-principles study of the structural, energetic, and electronic properties of single-layer graphene doped with boron and nitrogen atoms with varying doping concentrations and configurations is described in Reference [11]. The calculations in the paper indicate that the bandgap can be adjusted as required based on the doping concentration and the doping configuration. The modeling and design of a new flexible graphene-on-silicon Schottky junction solar cell with a power conversion efficiency >10% is described in Reference [12]. Reference [13] proposes asymmetric double-well potential on graphene as an electronic waveguide to confine the graphene electrons. The guided modes in this graphene waveguide are investigated using a modified transfer matrix method.

An overview of recent advances in electronic and optoelectronic devices based on 2D TMDCs is presented in Reference [14]. This review focuses on evaluating field-effect transistors (FETs), photovoltaic cells, light-emitting diodes (LEDs), photodetectors, lasers, and integrated circuits (ICs) using TMDCs. The review "Photonic Structure-Integrated Two-Dimensional Material Optoelectronics" offers an overview and evaluation of state-of-the-art of hybrid systems, where 2D material optoelectronics are integrated with photonic structures, especially plasmonic nanostructures, photonic waveguides, and crystals [15].

3. Future

While the potential of 2D materials in future flexible electronics and optoelectronics has been widely recognized by the scientific community, several major challenges still remain. The lack of effective methods to p- and n-dope 2D semiconductors has seriously restricted their device applications. The presence of a significant contact barrier between most 2D semiconductors and common electrode materials has so far limited the performance of 2D electronics and prevented the ultimate downscaling of the device dimensions. Techniques to grow high quality 2D materials with wafer-scale uniformity need to be developed to scale up the production of 2D electronic and optoelectronic devices. The low carrier mobility in 2D semiconductors appears to limit their advantages over the current technology. Future innovations that overcome these bottlenecks will likely lead to breakthroughs in 2D electronics and optoelectronics. New 2D materials with promising electronic and optoelectronic properties are also likely to emerge in the future.

Acknowledgments: First of all we would like to thank all researchers who submitted articles to this special issue for their excellent contributions. We are also grateful to all reviewers who helped in the evaluation of the manuscripts and made very valuable suggestions to improve the quality of contributions. We would like to acknowledge the editorial board of *Electronics*, who invited us to guest edit this special issue. We are also grateful to the *Electronics* Editorial Office staff who worked thoroughly to maintain the rigorous peer-review schedule and timely publication. Y.K.Y. is supported by National Science Foundation (Award number DMR-1261910) during the course of editing this special issue.

Author Contributions: Z.Z. and Y.K.Y. worked together in the whole editorial process of the special issue, 'Two-Dimensional Electronics and Optoelectronics', published by journal *Electronics*. Z.Z. drafted this editorial summary. Z.Z. and Y.K.Y. reviewed, edited and finalized the manuscript.

Conflicts of Interest: The authors declare no conflicts of interest.

References

1. Balandin, A.A.; Ghosh, S.; Bao, W.; Calizo, I.; Teweldebrhan, D.; Miao, F.; Lau, C.N. Superior Thermal Conductivity of Single-Layer Graphene. *Nano Lett.* **2008**, *8*, 902–907. [CrossRef] [PubMed]
2. Ghosh, S.; Calizo, I.; Teweldebrhan, D.; Pokatilov, E.P.; Nika, D.L.; Balandin, A.A.; Bao, W.; Miao, F.; Lau, C.N. Extremely high thermal conductivity of graphene: Prospects for thermal management applications in nanoelectronic circuits. *Appl. Phys. Lett.* **2008**, *92*, 151911. [CrossRef]
3. Seol, J.H.; Jo, I.; Moore, A.L.; Lindsay, L.; Aitken, Z.H.; Pettes, M.T.; Li, X.; Yao, Z.; Huang, R.; Broido, D.; et al. Two-Dimensional Phonon Transport in Supported Graphene. *Science* **2010**, *328*, 213–216. [CrossRef] [PubMed]
4. Bolotin, K.I.; Sikes, K.J.; Jiang, Z.; Klima, M.; Fudenberg, G.; Hone, J.; Kim, P.; Stormer, H.L. Ultrahigh electron mobility in suspended grapheme. *Solid State Commun.* **2008**, *146*, 351–355. [CrossRef]
5. Radisavljevic, B.; Radenovic, A.; Brivio, J.; Giacometti, V.; Kis, A. Single-layer MoS$_2$ transistors. *Nat. Nanotechnol.* **2011**, *6*, 147–150. [CrossRef] [PubMed]
6. Novoselov, K.S.; Jiang, D.; Schedin, F.; Booth, T.J.; Khotkevich, V.V.; Morozov, S.V.; Geim, A.K. Two-dimensional atomic crystals. *Proc. Natl. Acad. Sci. USA* **2005**, *102*, 10451–10453. [CrossRef] [PubMed]
7. Li, L.; Yu, Y.; Ye, G.J.; Ge, Q.; Ou, X.; Wu, H.; Feng, D.; Chen, X.H.; Zhang, Y. Black phosphorus field-effect transistors. *Nat. Nanotechnol.* **2014**, *9*, 372–377. [CrossRef] [PubMed]
8. Liu, H.; Neal, A.T.; Zhu, Z.; Luo, Z.; Xu, X.; Tománek, D.; Ye, P.D. Phosphorene: An Unexplored 2D Semiconductor with a High Hole Mobility. *ACS Nanotechnol.* **2014**, *8*, 4033–4041. [CrossRef] [PubMed]

9. Ghasemi, F.; Frisenda, R.; Dumcenco, D.; Kis, A.; Perez de Lara, D.; Castellanos-Gomez, A. High Throughput Characterization of Epitaxially Grown Single-Layer MoS$_2$. *Electronics* **2017**, *6*, 28. [CrossRef]
10. Browning, R.; Kuperman, N.; Moon, B.; Solanki, R. Atomic Layer Growth of InSe and Sb$_2$Se$_3$ Layered Semiconductors and Their Heterostructure. *Electronics* **2017**, *6*, 27. [CrossRef]
11. Varghese, S.; Swaminathan, S.; Singh, K.; Mittal, V. Energetic Stabilities, Structural and Electronic Properties of Monolayer Graphene Doped with Boron and Nitrogen Atoms. *Electronics* **2016**, *5*, 91. [CrossRef]
12. Xu, Y.; Ang, L. Guided Modes in a Double-Well Asymmetric Potential of a Graphene Waveguide. *Electronics* **2016**, *5*, 87. [CrossRef]
13. Dell'Olio, F.; Palmitessa, M.; Ciminelli, C. Modeling and Design of a New Flexible Graphene-on-Silicon Schottky Junction Solar Cell. *Electronics* **2016**, *5*, 73. [CrossRef]
14. Ye, M.; Zhang, D.; Yap, Y. Recent Advances in Electronic and Optoelectronic Devices Based on Two-Dimensional Transition Metal Dichalcogenides. *Electronics* **2017**, *6*, 43.
15. Wang, T.; Xu, Y.-Q. Photonic Structure-Integrated Two-Dimensional Material Optoelectronics. *Electronics* **2016**, *5*, 93. [CrossRef]

electronics

MDPI

Review

Photonic Structure-Integrated Two-Dimensional Material Optoelectronics

Tianjiao Wang [1] and Ya-Qiong Xu [1,2,*]

[1] Department of Electrical Engineering and Computer Science, Vanderbilt University, Nashville, TN 37235-1824, USA; tianjiao.wang@vanderbilt.edu
[2] Department of Physics and Astronomy, Vanderbilt University, Nashville, TN 37235-1807, USA
*   Correspondence: yaqiong.xu@vanderbilt.edu

Academic Editors: Yoke Khin Yap and Zhixian Zhou
Received: 25 October 2016; Accepted: 9 December 2016; Published: 20 December 2016

Abstract: The rapid development and unique properties of two-dimensional (2D) materials, such as graphene, phosphorene and transition metal dichalcogenides enable them to become intriguing candidates for future optoelectronic applications. To maximize the potential of 2D material-based optoelectronics, various photonic structures are integrated to form photonic structure/2D material hybrid systems so that the device performance can be manipulated in controllable ways. Here, we first introduce the photocurrent-generation mechanisms of 2D material-based optoelectronics and their performance. We then offer an overview and evaluation of the state-of-the-art of hybrid systems, where 2D material optoelectronics are integrated with photonic structures, especially plasmonic nanostructures, photonic waveguides and crystals. By combining with those photonic structures, the performance of 2D material optoelectronics can be further enhanced, and on the other side, a high-performance modulator can be achieved by electrostatically tuning 2D materials. Finally, 2D material-based photodetector can also become an efficient probe to learn the light-matter interactions of photonic structures. Those hybrid systems combine the advantages of 2D materials and photonic structures, providing further capacity for high-performance optoelectronics.

Keywords: two-dimensional materials; plasmonics; photonic crystals; optoelectronics

1. Introduction

Two-dimensional (2D) materials have attracted extensive attention since the last century due to the wealth of novel physical properties when charge and heat transport are confined to the direction perpendicular to a 2D plane. The isolation of graphene in 2004 by Geim and Novoselov first showed the possibility to obtain stable, single-atom layer 2D materials from their van der Waals solids [1]. In the past decade, graphene has been demonstrated to possess many outstanding merits not only for thermal and mechanical applications, but also for electronics and optoelectronics [2–10]. The great success of graphene encourages researchers to rediscover and restudy other 2D materials [11–15]. One of the most well-studied 2D material families is transition metal dichalcogenide (TMDC). TMDCs have a general chemical formula of MX_2, where M is a transition metal atom from groups IV, V, and VI (e.g., Mo, W), and X is a chalcogen atom (e.g., S, Se, Te) [16,17]. The variety of elements and the layer-dependence enable TMDCs to present a wide range of electrical, optical, chemical, thermal and mechanical properties [11,18–26]. Two distinctive features of TMDCs are strong excitonic effects and valley/spin-dependent properties. TMDCs process indirect bandgap for bulk crystals, while in the single layer limit, they become direct-bandgap semiconductors with gaps located at the K and the K' points [18,27–29]. Moreover, the broken in-plane inversion symmetry in monolayer give rise to valley-dependent optical and electrical properties [24,30–32]. Graphene analogues are another important type of 2D material, which includes hexagonal boron nitride (hBN) with a large bandgap up to 6 eV [6,33], anisotropic black phosphorus (BP) [15,34–39], and the most recent addition, boronphene [40]. The last category of 2D materials is transition metal oxide, including titania- and perovskite-oxides. Those oxide nanosheets have exhibited great potential for new capacitors and energy-storage devices [41].

Interestingly, 2D materials present several advantages over conventional three-dimensional (3D) materials for optoelectronics. First, although the innate thinness renders these materials almost transparent, their strong light-matter interactions enable decent single-pass absorption. For example, single-layer graphene absorbs 2.3% of vertically incident white light [42]. Monolayer molybdenum disulfide (MoS_2) absorbs around 10% at excitonic resonances [43]. Moreover, 2D materials cover a wide response spectral range from microwave to ultraviolet wavelengths. For instance, the zero bandgap makes graphene a potential candidate for optical applications over a broad spectral range. However, its semi-metallic nature prevents the realization of efficient optoelectronics due to a large dark current. TMDCs possess relatively larger bandgaps, enabling excellent on/off ratio, but limiting their performance in telecom-wavelength. BP has a layer-tunable direct bandgap ranging from 0.3 eV in bulk to 1.8–2.0 eV in monolayer [36], which covers from visible to mid-infrared spectral regions. Finally, the absence of dangling bonds makes 2D materials easy to be integrated with photonic structures or stacked together to form vertical van der Waals heterostructures [44].

Integrating external photonic structures with 2D material-based optoelectronics is a novel strategy to broaden the horizon of applications for 2D materials. Those metallic and dielectric photonic or nanophotonic structures can be precisely sculpted into various architectures so that light can be scattered, confined, refracted and processed in controllable ways [45–47]. Patterning 2D materials into arrays is a straightforward approach to form internal photonic structures. By sculpting 2D materials into plasmonic nanostructures like resonators or nanocavities, enhanced light absorption can be achieved [48–52]. Furthermore, external metallic plasmonic nanostructures enable light manipulation beyond the diffraction limit of light by the excitation of surface plasmons, which confines the electromagnetic field at the metal surface. Photonic waveguides provide another possibility to control light flows. With the assistance of waveguides, light can be side coupled into 2D material devices, leading to enhanced responsivity. The combination of photonic structures with 2D material-based optoelectronic devices provides the possibility to enhance the overall performance since special requirements at a certain location, polarization direction or wavelength can be achieved by choosing different photonic structures or tuning their geometric parameters. On the other hand, combining 2D material-based devices with classic photonic modulators offers an approach to tuning the modulator performance by electrostatically manipulating 2D materials. Moreover, the 2D material/photonic structures hybrid systems allow 2D material optoelectronics to become a probe to investigate light-matter interactions of photonic structures.

In this review, we will first briefly overview 2D material-based optoelectronics, especially photodetectors, and their physical mechanisms. For the purpose of this review, we broadly define 2D materials to be thin-layered films and those atomically thin layers are held together only by van der Waals force. Also, the total thicknesses of the thin films can vary from several angstroms to tens of nanometers. In the following, we will give a general discussion about the structure and performance of photonic structure/2D material hybrid systems, in which 2D optoelectronics are combined with different photonic structures, including plasmonic nanostructures, photonic waveguides and crystals. Next, we will present the possibility that 2D material optoelectronics can influence and even probe the properties of photonic structures. Finally, we will summarize the main conclusions of the review and describe the future opportunities in this promising field of research.

2. Two-Dimensional (2D) Material-Based Optoelectronics

In addition to the outstanding electronic properties of 2D materials, such as high mobility of graphene devices and high on/off ratio of TMDC-based transistors, their unique optical properties also attract researchers to learn their optoelectronic characterizations [10,17,53–59].

2.1. Photocurrent-Generation Mechanisms

Generally, the photoresponses of 2D material-based devices can be attributed to more than one mechanism when device structures vary or are investigated under different light illumination conditions. Here, several widely studied mechanisms in typical 2D material-based optoelectronics will be described, including photovoltaic effect, photo-thermoelectric effect, photo-bolometric effect, and hot electron injection. Figure 1 summarizes these mechanisms in terms of a typical 2D material-based phototransistor, in which a 2D material thin film bridges two metal electrodes. Here we plot semiconducting 2D material channels rather than semimetal channels to show more comprehensive circumstances. Additionally, photoconductive effect and photogating can also contribute to photocurrent generation [60].

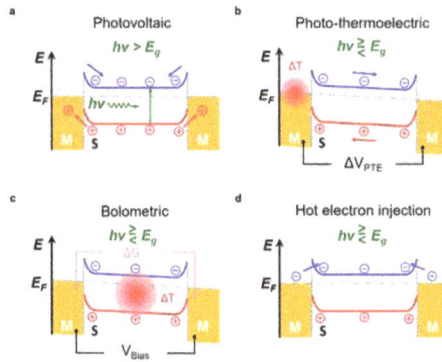

Figure 1. Schematic representation of photocurrent-generation mechanisms in semiconducting 2D materials. M and S indicate metal electrode contacts and 2D semiconductor channels, respectively. (a) Photon-excited electron-hole pairs (EHPs) separated by internal electric fields at metal-semiconductor Schottky barriers. Red shaded areas indicate an elevated temperature ΔT induced by laser heating, leading to (b) a voltage difference ΔV_{PTE} or (c) overall conductance change ΔG across the channel; (d) Carriers are injected from metal electrodes into a 2D material channel.

2.2. Photovoltaic Effect (PVE)

In early studies, the photocurrent generation in 2D materials is attributed to the separation of photon-excited electron-hole pairs (EHPs), usually at PN junctions or metal-semiconductor Schottky barriers, where internal electric fields force the photo-excited EHPs to be separated. This phenomenon is known as photovoltaic effect (PVE), which has been elaborated in traditional 3D semiconductor devices [61,62]. Here we note that photocurrent is defined to be $\Delta I_{pc} = I_{ds,illumination} - I_{ds,dark}$. As shown in Figure 1a, under illumination with photon energy larger than the bandgap of 2D materials, electrons in the valence band can be excited to the conduction band, producing photo-excited EHPs. With the assistance of internal electric fields, excess electrons and holes will be driven to opposite directions, leading to a light-generated current (I_L). When the circuit is open, the accumulation of carriers induces a voltage (open circuit voltage V_{oc}), which lowers the potential barrier and generates a forward bias diffusion current that balances the light-generated current. If the external bias is set to be zero, the photo-excited EHPs will be collected, generating a photocurrent (short circuit current I_{sc}). For typical graphene or 2D material-based phototransistors, PVE is thought to be dominant since Schottky barriers usually exist when 2D materials contact metal electrodes. By applying various external gate voltages on an MoS$_2$ device, as illustrated in Figure 2a, the conduction/valence band level of single-layer MoS$_2$ will be tuned accordingly [63]. Thus, the band alignment presents great difference under various gate voltages. Since the band alignment will directly affect the direction and value of the internal electric field, the sign and intensity of PVE-induced photocurrent will change significantly when the gate voltage is sweeping. The drain bias is also very important in terms of PVE since it provides a way to tune the lateral band alignment.

2.3. Photo-Thermoelectric Effect and Photo-Bolometric Effect

The thermal mechanisms can also drive photocurrent generation. In those cases, photo-thermoelectric effect (PTE) and photo-bolometric effect (PBE) can contribute to the photocurrent generation by inducing non-uniform heat and an overall conductance change of the 2D material channel under light illumination, respectively. Seebeck coefficient is defined as the ability to induce thermoelectric voltage in response to a temperature difference across the material and can usually be expressed through the Mott relation [64,65]:

$$S = \frac{\pi^2 k_B^2 T}{3e} \frac{1}{G} \frac{dG}{dE}\bigg|_{E=E_F} \tag{1}$$

where k_B is the Boltzmann constant, G is conductance, e is the electron charge, and E_F is Fermi energy. For a semiconductor channel that connected to two metal electrodes, the light-induced temperature increase (ΔT) at the metal electrodes will generate a photo-thermoelectric voltage across the channel as displayed in Figure 1b. PTE drives a current through the junction even without applying an external bias, and the current value is directly related to the channel conductance and the photo-thermoelectric voltage, which is determined by the Seebeck coefficient difference between the semiconductor channel and the metal electrodes, and the temperature change:

$$\Delta V_{PTE} = (S_{semiconductor} - S_{metal})\Delta T \tag{2}$$

The Seebeck coefficients of pure metals are typically in the order of 1 μV/K. However, for 2D materials (such as graphene, TMDCs and BP), the Seebeck coefficient ranges from several to thousands μV/K [66–73]. On the other hand, as shown in Figure 1c, the local heat induced by photon absorption can modify the resistance of the channel. This light-induced conductance will lead to photocurrent under an external bias by PBE [74,75]. The PBE-induced photocurrent can be presented by the following equation:

$$I_{PBE} = \Delta G V_D \tag{3}$$

where ΔG is the conductance change and V_D is the external bias. From Equation (3), we find that the PBE-induced photocurrent is predicted to present a linear relationship to the applied bias. In previous investigations, PTE and PBE have been found to play indispensable roles in 2D material-based devices. Although PTE-induced photocurrent is negligible in the case that the illumination is focused within a uniform semiconducting channel, PTE effect can be observed for a nonuniformity doping, for instance, a graphene junction consisting of bilayer and monolayer [66]. A report by Gabor et al. further demonstrated the PTE in graphene by studying a dual-gated graphene PN junction, as presented in Figure 2b [76]. By changing the biases that were applied to the top and bottom gates, the doping level of the two sides of the junction can be precisely and independently controlled. The photoresponses exhibit six-fold polarity variation, which is attributed to the changes of the Seebeck coefficient of graphene when the gate bias varies. Additionally, PTE effect can be observed at 2D material/metal interface. A study about monolayer MoS_2 transistors has reported that the photoresponse at MoS_2/metal junction is dominant by PTE and demonstrated large Seebeck coefficient for MoS_2 under an external electric field [71]. On the other hand, PBE has been comprehensively studied for carbon nanotube devices, in which photoresponse shows linear negative correlation to the external bias [75]. PBE-induced photocurrents have also been observed by Freitag et al. when the graphene channel is heavily doped (Figure 2c) [77], in which overall current drops under light excitation since the temperature increase leads to a reduction in carrier mobility. Similar phenomena have also been detected in heavily doped BP devices [72].

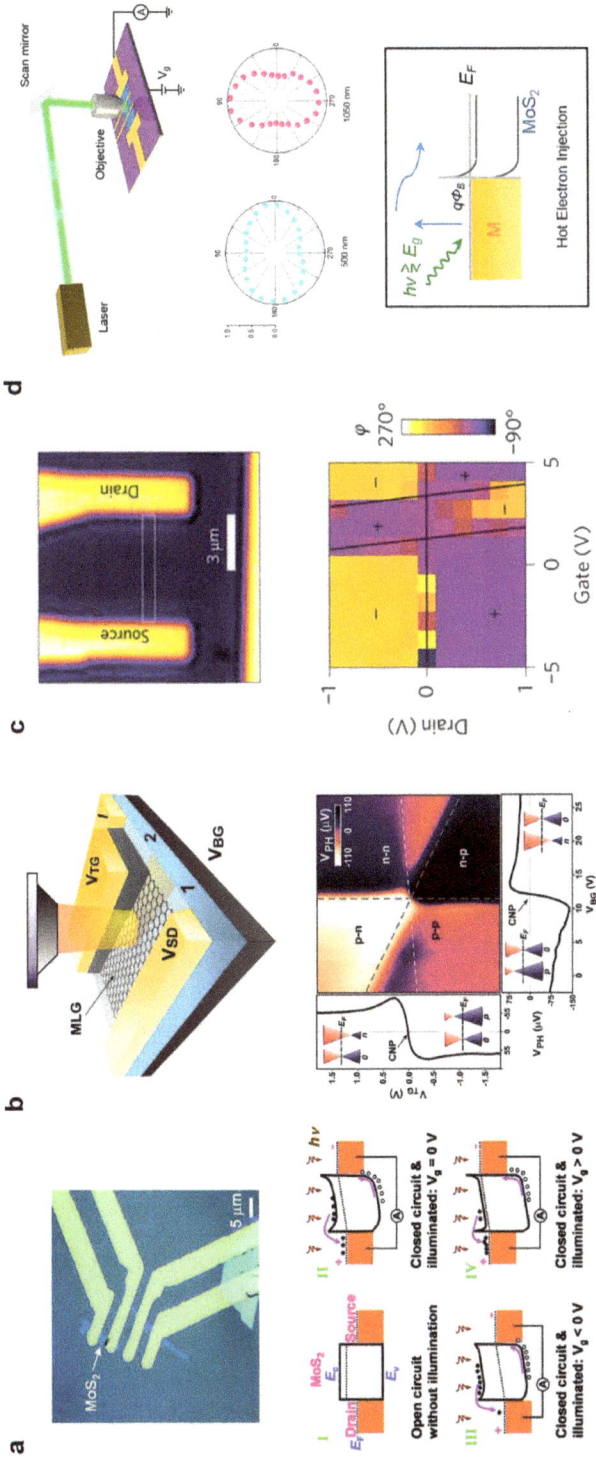

Figure 2. Photocurrent generation in 2D materials. (**a**) Optical image of monolayer MoS$_2$-based field effect transistor (top) and band diagrams under different illumination and bias conditions (bottom), reproduced with permission from [63], ACS, 2012; (**b**) 3D schematic of a dual-gated graphene phototransistor (top) and six-fold photovoltage pattern occurs while the top and bottom gate biases change (bottom), reproduced with permission from [76], AAAS, 2011; (**c**) Reflection image of a strongly doped graphene phototransistor (top) and bolometric photocurrent observed to exhibit opposite polarities to applied biases (bottom), reproduced with permission from [77], Nature Publishing Group, 2012; (**d**) A four-electrode MoS$_2$ phototransistor (top), photocurrent signals that show different anisotropic behaviors with photo energy above/below the MoS$_2$ bandgap (middle), and hot electron injection that is proposed to be dominant when the photon energy is below the band gap of MoS$_2$ (bottom), reproduced with permission from [78], ACS, 2015.

2.4. Hot Electron Injection

With further investigation of 2D material optoelectronics, hot electron injection has been demonstrated as a novel photocurrent-generation mechanism [78]. It is well-known that photon-excited hot electrons in metal electrodes can cross over the Schottky barrier and be injected into the semiconductor channel. The injection yield of electrons Y follows the Fowler equation:

$$Y \sim \frac{1}{8E_F} \frac{(\hbar\omega - \phi_B)^2}{\hbar\omega} \tag{4}$$

where E_F is the Fermi energy, $\hbar$ is the reduced plank constant, ω is the incident light frequency, and ϕ_B is the Schottky barrier. Hot electron injection has been reported to be indispensable in TMDC-based devices, such as a MoS$_2$-based phototransistor (Figure 2d) [78]. Anisotropic photocurrent response has been detected under illumination with photon energy below the MoS$_2$ bandgap. More importantly, the anisotropic ratios are found to be related to the shape of Au electrodes, indicating the hot electron contribution from metal electrodes. Additionally, hot electron injection also plays an important role when 2D materials directly contact with silicon photonic crystals and metal plasmonic nanostructures [79].

For a practical device, the situation is more complex, usually present a combined effect of different mechanisms [76]. Therefore, many strategies are adopted to learning the photocurrent-generation mechanisms for various devices. Unlike graphene, TMDCs present slightly complicated scenarios due to the existence of appreciable bandgaps. For the most well-known Mo- and W-based TMDCs, bandgaps are typically ranged from 1 to 2 eV. Wavelength-dependence and polarization-dependence measurements are conducted to further probe the underlying physical mechanisms. From those studies, various photocurrent-generation mechanisms are reported under different illumination conditions [21,78,80]. Taking MoS$_2$ as an example, when the photon energy is high enough to generate photo-excited EHPs, large photocurrent signals can be observed at junction areas because of PVE. However, measurable photocurrent signals can also be detected when incident photon energy is below the bandgap of MoS$_2$. In this case, PTE can explain the photocurrent generation at metal electrode areas. Moreover, at metal-semiconductor junctions, photocurrent signals are detected to exhibit polarization dependency perpendicular to the metal edge, which is attributed to the hot electron injection [78].

As the investigation of photocurrent-generation mechanisms in 2D optoelectronics goes deeper, 2D material-based devices have also been demonstrated to exhibit elevated performance. Over the past decade, it has been shown that most graphene-based phototransistors exhibit photoresponsivity around 10 mA/W [9,81]. By introducing electron trapping centres, high broadband photoresponse is demonstrated to achieve 8.61 A/W [8]. Moreover, due to the high carrier mobility and high carrier-saturation velocity, graphene-based optoelectronics are reported to operate at speed up to tens of gigahertz [81]. First reported single-layer MoS$_2$ phototransistors reach responsivity to 7.5 mA/W with 50 V gate bias, and exhibit stable response time within 50 ms [63]. An ultrasensitive monolayer MoS$_2$ photodetector was reported later to show a maximum external photoresponsivity of 880 A/W at 561 nm [21]. Other TMDCs, such as MoSe$_2$, WS$_2$ and WSe$_2$, have also been employed for optoelectronics, and achieve photoresponsivities ranging from tens of A/Ws to hundreds [82]. It must be mentioned that an electrostatically defined PN junction has been realized by locally gating the WSe$_2$ thin film [19]. BP is a new addition to the 2D family, and it attracts tremendous attention in the optoelectronic field due to its unique optical properties, especially its anisotropic nature and layer-dependent bandgaps, which covers the visible to mid-infrared spectral range. BP-based photodetectors are reported to reach photoresponsivity up to 4.8 mA/W [39]. Additionally, due to the directional dependence of the interband transition strength in the anisotropic band structure of BP, polarized photocurrent responses are observed in BP phototransistors via PVE [83]. On the other hand, PTE and PBE-induced photocurrent generation is also considered to be important, especially in metallic or doped multilayer BP phototransistors [72].

In addition to the 2D material-based optoelectronics in the above discussions, 2D material-based heterostructures have been reported to show excellent performance for optoelectronic applications in the previous literatures [84–89]. By stacking different 2D semiconducting crystals on top of each other with van der Waals-like forces, those heterostructures are expected to show combined functionality of the individual layers and new phenomena at the interface. For example, a maximum photodetection responsivity of 418 mA/W at the wavelength of 633 nm has been achieved by stacking p-type BP with n-type MoS_2 to form ultrathin PN diodes [90]. The photocurrent signals at the PN junction are much stronger than photoresponse at the metal-semiconductor contacts due to the large band mismatch at the heterojunction. The polarization direction of the photoresponse at the junction is affected by either BP or MoS_2 channel, depending on the illumination conditions [91]. Moreover, 2D material-based heterostructures allow the possibility to obtain high quality devices by encapsulating a 2D material thin film with 2D insulators such as hBN sheets to form a sandwich-like structure. The hBN sheets provide an ultra-smooth surface for the inner graphene film and protect the graphene film from direct exposure to air or contact with substrates. These devices exhibit room-temperature mobility up to 140,000 cm^2/Vs which is near the theoretical phonon-scattering limit [6,92].

3. Photonic Structure-Integrated 2D Material Optoelectronics

The integration of conventional plasmonic nanostructures, waveguides and photonic crystals with 2D material-based optoelectronics could be beneficial for both fields of investigation. First, the introduction of photonic structures into 2D materials can significantly enhance their photoresponsivity [93–106]. On the other hand, by electrostatically tuning the Fermi level of 2D materials, with which photonic structures like waveguides, resonators and cavities are integrated, their modulation performance can be improved [107–113]. Table 1 summarizes different photonic structure-integrated 2D material optoelectronics.

Table 1. Photonic structure-integrated 2D material hybrids for optoelectronics.

Application	Photonic Structures		2D Material	Performance	Ref.
Photodetector	Plasmonic Nanostructure	Au Particle	MoS_2	Photoresponsivity: 200%–300% enhancement	[99]
		Au Antenna Array	MoS_2	Photoresponsivity: 5.2 A/W Photogain: 10^5	[100]
		Au Particle	Graphene	Photoresponsivity: 1500% enhancement	[101]
		Ti/Au nanostructures	Graphene	Photoresponsivity: 2000% enhancement	[102]
	Waveguide	Si Waveguide	BP	Photoresponsivity: 657 mA/W response rate: ~3 GHz	[103]
		Si Waveguide	Graphene	Photoresponsivity: 0.1 A/W response rate: ~20 GHz	[104]
		Si Waveguide	Graphene	Response wavelength range: all optical communication band response rate: ~18 GHz	[105]
		Suspended Waveguide	Graphene	Photoresponsivity: 0.13 A/W	[106]
Modulator	Waveguide	Si Waveguide	Graphene	Modulation Depth: 3 dB operation speed: over 1 GHz bandwidth: 1.35–1.6 μm	[107]
		Si Microring Resonator	Graphene	Modulation Depth: 12.5 dB	[108]
	Photonic Crystal	2D Photonic Crystal Nanocavity	Graphene	Modulation Depth: 6 dB	[110]

3.1. Photodetector

The 2D materials have remarkably high single-pass absorption. However, most light will still transmit through the thin layers as the total thickness is below tens of nanometers. Therefore, the light-matter interactions must be further enhanced before it can be used in practical applications, especially for light detection. The combinations of plasmonic nanostructures with 2D materials are first studied to enhance light-matter interactions. Placing metallic objects by an ordered array with dimension and pitch in the order of the excitation wavelength is a classic way to form plasmonic nanostructures. Either stacking 2D materials on top of those plasmonic nanostructures, or patterning plasmonic nanostructures on top of 2D materials can increase the light absorption at certain wavelengths [93–97]. A significant enhancement ~65% of photoluminescence intensity are reported for a monolayer MoS_2-coated gold nanoantennas system [98]. Similar phenomena are observed for 2D material base optoelectronics. A large enhancement of photocurrent response can be obtained by coupling few-layer MoS_2 with Au plasmonic nanostructure arrays [99]. Depositing 4 nm thick Au nanoparticles sparsely onto few-layer MoS_2 phototransistors leads to a two-fold increase in the photocurrent response. To further enhance the photocurrent response, well-defined periodic Au nanoarrays are synthesized on few-layer MoS_2. Under this circumstance, a three-fold enhancement is achieved, which is attributed to the light trap near the Au nanoplates. On the other hand, a direct contact between 2D materials and metals enables hot electron injection over Schottky barriers [100]. A bilayer MoS_2 film has been integrated with a plasmonic antenna array as shown in Figure 3a. With the assistance of hot electron injection, a photoresponsivity of 5.2 A/W has been achieved under 1070 nm illumination, and a photogain is estimated to be ~10^5. No noticeable photocurrent signals are observed if an Al_2O_3 barrier exists between the MoS_2 film and Au array, suggesting that hot electron injection indeed plays a dominant role in the photocurrent generation. Similar phenomena are observed for graphene. In Figure 3b, gold nanoparticles are transferred onto pre-prepared graphene transistors [101]. Here, the plasmonic nanostructures can be treated as subwavelength scattering sources and nanoantennas to enhance the optical detection and photoresponse at selected plasmon resonance frequencies. For this metallic plasmonic nanostructure-integrated graphene photodetector, a huge enhancement of photocurrent and external quantum efficiency up to 1500% has been achieved. By employing plasmonic nanostructures of different geometries, graphene photodetectors present not only enhanced responsivity but also selectivity of polarization [102]. As shown in Figure 3c, by combining graphene with plasmonic nanostructures, the efficiency of graphene-based photodetectors can be increased by up to 20 times. Moreover, for the special finger plasmonic nanostructure, the photocurrent signals show strong polarization dependency.

Photonic waveguides are another type of photonic structures that have been extensively studied to be combined with 2D optoelectronics to achieve a high-performance photodetector. Instead of top coupling, the incident light can be guided into the 2D photodetector by a waveguide through side coupling in order to achieve high photoresponsivity and fast response rate. In a recent report, a silicon waveguide-integrated BP photodetector has been demonstrated [103]. In comparison with graphene, the narrow bandgap nature allows BP to operate in the near-infrared band and, at the same time, preserve low dark current. As shown in Figure 4a, a silicon waveguide is planarized by SiO_2 to accept an exfoliated BP thin layer. A graphene gate is then fabricated to control the doping of BP to optimize the device performance. The 11.5 nm thick BP device has obtained an intrinsic responsivity up to 135 mA/W, and an even higher responsivity up to 657 mA/W has been observed for 100 nm thick BP. Also, a high response speed ~3 GHz has been achieved while the dark current is limited to 220 nA. Graphene/silicon waveguide hybrid systems have also been demonstrated to present high performance. As shown in Figure 4b [104], a silicon waveguide is backfilled with SiO_2 and then planarized. A 10 nm SiO_2 layer is subsequently deposited on the planarized surface to electrically isolate the graphene layer from the underlying silicon structure. Two metal electrodes are placed asymmetrically to the waveguide to maximize the photocurrent collection. Without waveguide coupling, photocurrent signals are observed at the position of the waveguide at a level of 2.6×10^{-4} A/W under normal light incidence. However, when the optical waveguide mode couples to the graphene layer through the evanescent field to enhance the optical absorption, the detector achieves a photoresponsivity exceeding 0.1 A/W with a nearly uniform response between 1450 and 1590 nm. Moreover, response rates exceeding 20 GHz are observed under zero bias operation. A similar silicon waveguide-integrated graphene photodetector (Figure 4c) is reported by Pospischil et al. to present ultra-wideband response covering all optical communication band with a high speed ~18 GHz [105]. In order to improve the performance of the graphene/silicon waveguide hybrids, suspended membrane silicon waveguides have been combined with graphene to form heterojunctions as displayed in Figure 4d [106]. The waveguide enables absorption of the evanescent light that propagates parallel to the graphene sheet, resulting in a responsivity as high as 0.13 A/W at a 1.5 V bias for 2.75 μm light at room temperature. Additionally, to explain the photocurrent dependence on bias polarity, direct transitions in graphene are proposed when photon energy is larger than the two-fold energy difference between the Fermi level and Dirac point. On the other hand, when the photon energy goes below this energy difference, indirect transitions will become dominant.

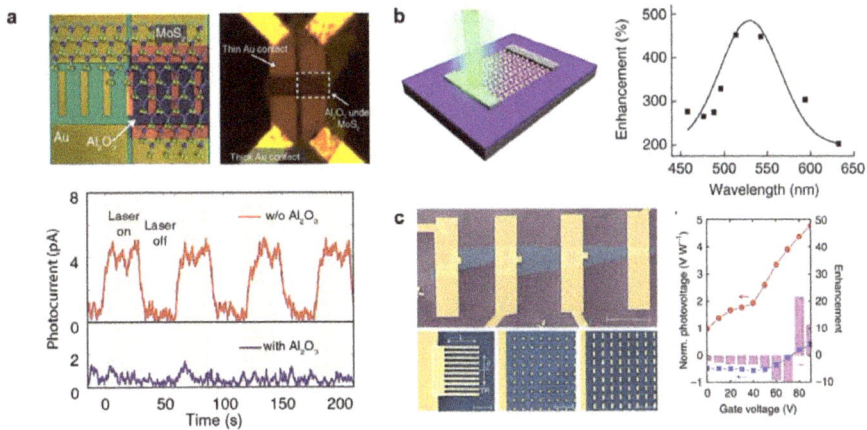

Figure 3. Plasmonic nanostructure-integrated 2D material photodetectors. (**a**) Schematic diagram (left-top) and optical image (right-top) of a device, in which MoS$_2$ is in direct contact with Au on the left side, while a 10 nm thick Al$_2$O$_3$ film is placed between MoS$_2$ and Au on the right side. Photocurrent measured from the left (upper panel) and the right (lower panel) subdevice (bottom), reproduced with permission from [100], ACS, 2015; (**b**) 3D schematic of a graphene transistor underneath gold nanoparticles (left) and photoresponse enhancement as a function of wavelength (right), reproduced with permission from [101], Nature Publishing Group, 2011; (**c**) Optical image of graphene phototransistors with different plasmonic nanostructures (left) and photovoltages at various gate biases (right), reproduced with permission from [102], Nature Publishing Group, 2011.

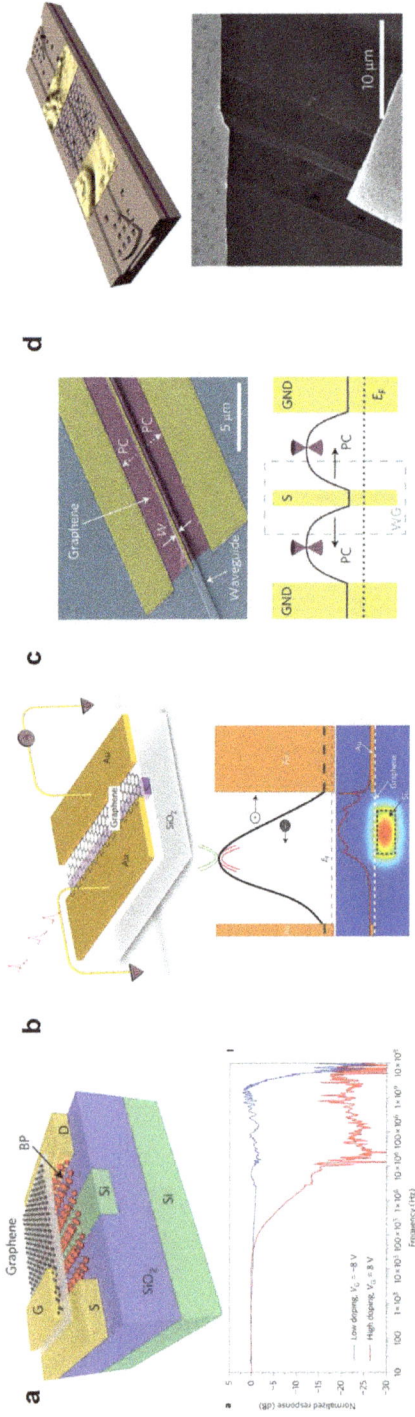

Figure 4. Waveguide-integrated 2D material photodetectors (**a**) 3D illustration of a silicon waveguide-integrated BP photodetector, featuring a few-layer graphene top-gate (top) and broadband frequency response of the BP photodetector with low or high doping (bottom), reproduced with permission from [103], Nature Publishing Group, 2013; (**b**) Schematic of a silicon waveguide-integrated graphene photodetector (top) and band diagram that shows photocurrent-generation mechanisms (bottom), reproduced with permission from [104], Nature Publishing Group, 2013; (**c**) Scanning electron micrograph (SEM) image of a silicon nanobeam-integrated graphene photodetector (top) and band diagram along the lateral direction (bottom), reproduced with permission from [105], Nature Publishing Group, 2013; (**d**) 3D schematic of a graphene-silicon heterostructure waveguide photodetector (top) and SEM image (bottom), reproduced with permission from [106], Nature Publishing Group, 2013.

3.2. Electro-Optic Modulator

The combination of photonic structures like waveguides, resonators or cavities with 2D materials enables the achievement of electro-optic modulator by electrostatically tuning 2D materials. A graphene-based broadband optical modulator is fabricated by placing a 250 nm-thick Si bus waveguide underneath a graphene sheet as illustrated in Figure 5a [107]. A 7 nm-thick Al_2O_3 spacer is placed between Si and graphene. The integration introduces several performance improvements: graphene provides strong light-matter interactions and broadband response covering telecommunication bandwidth, as well as mid- and far-infrared spectral regions. Furthermore, modulation of the guided light at frequencies over 1 GHz is demonstrated by electrically tuning the Fermi level of the graphene. Ring-shaped silicon microring resonators have also been reported to be combined with the graphene transistor as displayed in Figure 5b [108]. By tuning coverage lengths and electrostatic biases of graphene, a high extinction ratio of 12.8 dB is achieved. Photonic crystals are another extensively studied field to form various optical modulators such as resonators and cavities as the flow of light can be precisely molded. Photonic crystals are analogous to crystals where the unit structure is periodically duplicated [44–46]. When the periodicity in these structures approaches the wavelength of light, a photonic bandgap, in which the light propagation is forbidden, will appear. Under this circumstance, a pre-designed break in periodicity will introduce confinement of light. Photonic crystal/2D material hybrid systems are also reported in previous literatures [108–113]. Gan et al. have employed linear three-hole defect cavities in air-suspended 2D photonic crystals with graphene thin films transferred to the top [109]. The reflection attenuation and the increase of Raman signal intensity have demonstrated the local enhancement of light-matter interactions. Photonic crystals have also been reported to be integrated with 2D material phototransistors [110,111]. As shown in Figure 5c, optical cavities are fabricated by introducing point defects into 2D silicon photonic crystals [110]. The electrostatic properties of graphene are tuned by either source-drain biases or an ion-gel gate; as a result, a ~2 nm change in the cavity resonance line width and almost 400% (6 dB) modulation in resonance reflectivity are observed. Similar phenomena are reported for line defect photonic crystals [111].

In addition to the above discussions, the 2D optoelectronics also offer the possibilities to investigate the light modulation properties of photonic crystals considering the direct interaction between them [79]. Even photonic crystals have been extensively learned; their light manipulation manners can only be simulated by finite-difference time-domain (FDTD) simulations and detected by near-field scanning optical microscopy [114–117]. As described in Figure 4b, a graphene photodetector successfully detects a photocurrent strip on top of a silicon waveguide, which is attributed to the enhanced photocarrier density controlled by the waveguide. As a result, the photocurrent signals collected through the graphene photodetector directly visualize the light manipulation manners of the underneath silicon waveguide [104]. Recently, by directly contacting a featured silicon nanobeam with a BP photodetector, the spatially resolved photocurrent measurement is demonstrated to be an efficient approach to learn light-scattering properties of the silicon waveguide [79]. The direct contact between BP and the silicon waveguide enable photocarriers generated in the silicon to be injected into the BP. Since the density of photocarriers is related to the light distribution controlled by the waveguide, the spatial photocurrent signals detected by the BP photodetector reflect the light-scattering properties of the silicon waveguide, which shows strong dependence on both wavelength and polarization direction which are in good agreement with the prediction of FDTD.

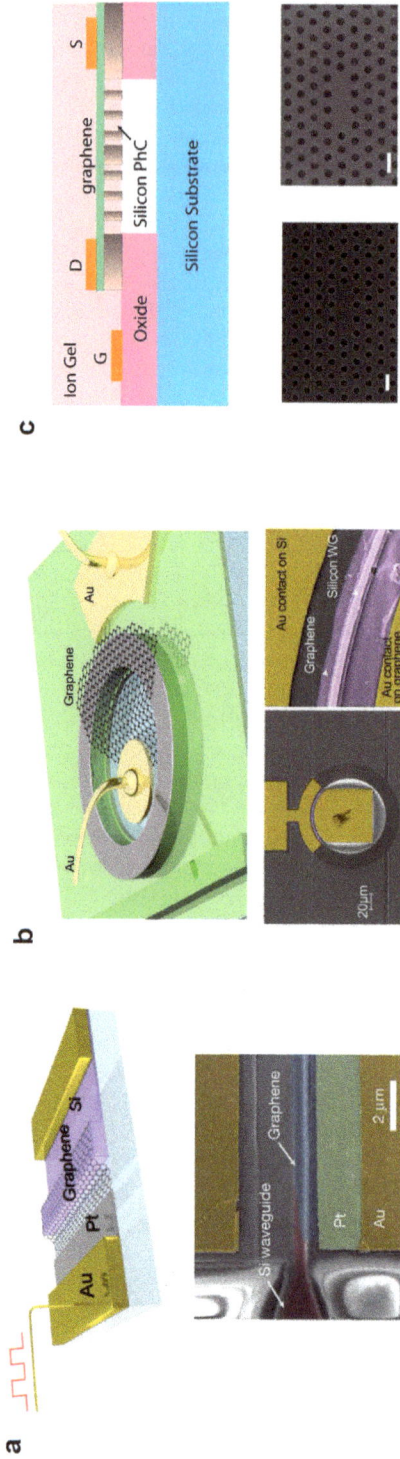

Figure 5. A 2D electronics-integrated optical modulator: (**a**) Schematic of a high-bandwidth graphene-integrated silicon waveguide modulator (top) and top view of the structure (bottom), reproduced with permission from [107], Nature Publishing Group, 2011; (**b**) Schematic of a graphene-silicon microring device (top), false-color SEM image of the fabricated device (left bottom), and a zoom-in SEM image of the bended silicon waveguide covered by graphene (right bottom), reproduced with permission from [108], ACS, 2015; (**c**) Schematic of the photonic crystal cavity–graphene device. SEM image of the fabricated cavities without (left bottom), and with graphene (right bottom), reproduced with permission from [110], ACS, 2013.

4. Summary and Outlook

In this review, we summarize the current state-of-the-art in two-dimensional (2D) material-based optoelectronics, especially their integration with photonic structures. Several photocurrent-generation mechanisms, which have been reported in 2D material-based optoelectronics, are first introduced. Photonic structures, such as plasmonic nanostructures, photonic waveguides and crystals are subsequently described. For plasmonic nanostructures in 2D devices, metallic objects are placed on top of 2D materials as an ordered array with dimension and pitch in the order of the excitation wavelength. Those plasmonic nanostructures enable enhanced light absorption at certain wavelength ranges. Photonic waveguides are another type of nanoarchitectures to manipulate light flows. With the assistance of photonic crystals, those 2D material-based devices can achieve response with a high rate and efficiency. Besides, those combinations also make it possible to improve the performance of optical modulators by electrostatically manipulating 2D materials and to probe light-matter interactions of photonic crystals by scanning photocurrent measurements of 2D material-based optoelectronics. The study of photonic structure-integrated 2D material optoelectronics opens the avenues to engineer optoelectronics to meet versatile requirements for future applications.

Acknowledgments: This work was supported by the National Science Foundation (ECCS-1055852 and CBET-1264982).

Author Contributions: Tianjiao Wang and Yaqiong Xu wrote the paper.

Conflicts of Interest: The authors declare no conflict of interest.

References

1. Geim, A.K.; Novoselov, K.S. The rise of graphene. *Nat. Mater.* **2007**, *6*, 183–191. [CrossRef] [PubMed]
2. Schwierz, F. Graphene transistors. *Nat. Nanotechnol.* **2010**, *5*, 487–496. [CrossRef] [PubMed]
3. Liao, L.; Lin, Y.-C.; Bao, M.; Cheng, R.; Bai, J.; Liu, Y.; Qu, Y.; Wang, K.L.; Huang, Y.; Duan, X. High-speed graphene transistors with a self-aligned nanowire gate. *Nature* **2010**, *467*, 305–308. [CrossRef] [PubMed]
4. Wu, Y.; Lin, Y.-M.; Bol, A.A.; Jenkins, K.A.; Xia, F.; Farmer, D.B.; Zhu, Y.; Avouris, P. High-frequency, scaled graphene transistors on diamond-like carbon. *Nature* **2011**, *472*, 74–78. [CrossRef] [PubMed]
5. Lin, Y.M.; Dimitrakopoulos, C.; Jenkins, K.A.; Farmer, D.B.; Chiu, H.Y.; Grill, A.; Avouris, P. 100-ghz transistors from wafer-scale epitaxial graphene. *Science* **2010**, *327*, 662. [CrossRef] [PubMed]
6. Dean, C.R.; Young, A.F.; Meric, I.; Lee, C.; Wang, L.; Sorgenfrei, S.; Watanabe, K.; Taniguchi, T.; Kim, P.; Shepard, K.L.; et al. Boron nitride substrates for high-quality graphene electronics. *Nat. Nanotechnol.* **2010**, *5*, 722–726. [CrossRef]
7. Britnell, L.; Gorbachev, R.V.; Geim, A.K.; Ponomarenko, L.A.; Mishchenko, A.; Greenaway, M.T.; Fromhold, T.M.; Novoselov, K.S.; Eaves, L. Resonant tunnelling and negative differential conductance in graphene transistors. *Nat. Commun.* **2013**, *4*, 1794. [CrossRef] [PubMed]
8. Zhang, B.Y.; Liu, T.; Meng, B.; Li, X.; Liang, G.; Hu, X.; Wang, Q.J. Broadband high photoresponse from pure monolayer graphene photodetector. *Nat. Commun.* **2013**, *4*, 1811. [CrossRef] [PubMed]
9. Xia, F.; Mueller, T.; Lin, Y.-m.; Valdes-Garcia, A.; Avouris, P. Ultrafast graphene photodetector. *Nat. Nanotechnol.* **2009**, *4*, 839–843. [CrossRef] [PubMed]
10. Bonaccorso, F.; Sun, Z.; Hasan, T.; Ferrari, A.C. Graphene photonics and optoelectronics. *Nat. Photonics* **2010**, *4*, 611–622. [CrossRef]
11. Radisavljevic, B.; Radenovic, A.; Brivio, J.; Giacometti, V.; Kis, A. Single-layer MoS_2 transistors. *Nat. Nanotechnol.* **2011**, *6*, 147–150. [CrossRef] [PubMed]
12. Larentis, S.; Fallahazad, B.; Tutuc, E. Field-effect transistors and intrinsic mobility in ultra-thin $MoSe_2$ layers. *Appl. Phys. Lett.* **2012**, *101*, 223104. [CrossRef]
13. Ramakrishna Matte, H.S.S.; Gomathi, A.; Manna, A.K.; Late, D.J.; Datta, R.; Pati, S.K.; Rao, C.N.R. MoS_2 and WS_2 analogues of graphene. *Angew. Chem.* **2010**, *122*, 4153–4156. [CrossRef]
14. Fang, H.; Chuang, S.; Chang, T.C.; Takei, K.; Takahashi, T.; Javey, A. High-performance single layered WSe_2 p-FETs with chemically doped contacts. *Nano Lett.* **2012**, *12*, 3788–3792. [CrossRef] [PubMed]

15. Xia, F.; Wang, H.; Jia, Y. Rediscovering black phosphorus as an anisotropic layered material for optoelectronics and electronics. *Nat. Commun.* **2014**, *5*, 4458. [CrossRef] [PubMed]
16. Wang, Q.H.; Kalantar-Zadeh, K.; Kis, A.; Coleman, J.N.; Strano, M.S. Electronics and optoelectronics of two-dimensional transition metal dichalcogenides. *Nat. Nanotechnol.* **2012**, *7*, 699–712. [CrossRef] [PubMed]
17. Mak, K.F.; Shan, J. Photonics and optoelectronics of 2d semiconductor transition metal dichalcogenides. *Nat. Photonics* **2016**, *10*, 216–226. [CrossRef]
18. Splendiani, A.; Sun, L.; Zhang, Y.; Li, T.; Kim, J.; Chim, C.-Y.; Galli, G.; Wang, F. Emerging photoluminescence in monolayer MoS$_2$. *Nano Lett.* **2010**, *10*, 1271–1275. [CrossRef] [PubMed]
19. Ross, J.S.; Klement, P.; Jones, A.M.; Ghimire, N.J.; Yan, J.; Mandrus, D.G.; Taniguchi, T.; Watanabe, K.; Kitamura, K.; Yao, W.; et al. Electrically tunable excitonic light-emitting diodes based on monolayer WSe$_2$ p-n junctions. *Nat. Nanotechnol.* **2014**, *9*, 268–272. [CrossRef] [PubMed]
20. Lee, C.; Yan, H.; Brus, L.E.; Heinz, T.F.; Hone, J.; Ryu, S. Anomalous lattice vibrations of single- and few-layer MoS$_2$. *ACS Nano* **2010**, *4*, 2695–2700. [CrossRef] [PubMed]
21. Lopez-Sanchez, O.; Lembke, D.; Kayci, M.; Radenovic, A.; Kis, A. Ultrasensitive photodetectors based on monolayer MoS$_2$. *Nat. Nano* **2013**, *8*, 497–501. [CrossRef] [PubMed]
22. Mak, K.F.; He, K.; Shan, J.; Heinz, T.F. Control of valley polarization in monolayer MoS$_2$ by optical helicity. *Nat. Nanotechnol.* **2012**, *7*, 494–498. [CrossRef] [PubMed]
23. Zeng, H.; Dai, J.; Yao, W.; Xiao, D.; Cui, X. Valley polarization in MoS$_2$ monolayers by optical pumping. *Nat. Nanotechnol.* **2012**, *7*, 490–493. [CrossRef] [PubMed]
24. Jones, A.M.; Yu, H.; Ghimire, N.J.; Wu, S.; Aivazian, G.; Ross, J.S.; Zhao, B.; Yan, J.; Mandrus, D.G.; Xiao, D.; et al. Optical generation of excitonic valley coherence in monolayer WSe$_2$. *Nat. Nanotechnol.* **2013**, *8*, 634–638. [CrossRef] [PubMed]
25. Chiritescu, C.; Cahill, D.G.; Nguyen, N.; Johnson, D.; Bodapati, A.; Keblinski, P.; Zschack, P. Ultralow thermal conductivity in disordered, layered WSe$_2$ crystals. *Science* **2007**, *315*, 351–353. [CrossRef] [PubMed]
26. Jariwala, D.; Sangwan, V.K.; Lauhon, L.J.; Marks, T.J.; Hersam, M.C. Emerging device applications for semiconducting two-dimensional transition metal dichalcogenides. *ACS Nano* **2014**, *8*, 1102–1120. [CrossRef] [PubMed]
27. Li, T.; Galli, G. Electronic properties of MoS$_2$ nanoparticles. *J. Phys. Chem. C* **2007**, *111*, 16192–16196. [CrossRef]
28. Cheiwchanchamnangij, T.; Lambrecht, W.R.L. Quasiparticle band structure calculation of monolayer, bilayer, and bulk MoS$_2$. *Phys. Rev. B* **2012**, *85*, 205302. [CrossRef]
29. Mak, K.F.; Lee, C.; Hone, J.; Shan, J.; Heinz, T.F. Atomically thin MoS$_2$: A new direct-gap semiconductor. *Phys. Rev. Lett.* **2010**, *105*, 136805. [CrossRef] [PubMed]
30. Xiao, D.; Liu, G.-B.; Feng, W.; Xu, X.; Yao, W. Coupled spin and valley physics in monolayers of MoS$_2$ and other group-VI dichalcogenides. *Phys. Rev. Lett.* **2012**, *108*, 196802. [CrossRef] [PubMed]
31. Cao, T.; Wang, G.; Han, W.; Ye, H.; Zhu, C.; Shi, J.; Niu, Q.; Tan, P.; Wang, E.; Liu, B.; et al. Valley-selective circular dichroism of monolayer molybdenum disulphide. *Nat. Commun.* **2012**, *3*, 887. [CrossRef] [PubMed]
32. Mak, K.F.; McGill, K.L.; Park, J.; McEuen, P.L. The valley hall effect in MoS$_2$ transistors. *Science* **2014**, *344*, 1489–1492. [CrossRef] [PubMed]
33. Shi, Y.; Hamsen, C.; Jia, X.; Kim, K.K.; Reina, A.; Hofmann, M.; Hsu, A.L.; Zhang, K.; Li, H.; Juang, Z.-Y.; et al. Synthesis of few-layer hexagonal boron nitride thin film by chemical vapor deposition. *Nano Lett.* **2010**, *10*, 4134–4139. [CrossRef] [PubMed]
34. Li, L.; Yu, Y.; Ye, G.J.; Ge, Q.; Ou, X.; Wu, H.; Feng, D.; Chen, X.H.; Zhang, Y. Black phosphorus field-effect transistors. *Nat. Nanotechnol.* **2014**, *9*, 372–377. [CrossRef] [PubMed]
35. Qiao, J.; Kong, X.; Hu, Z.-X.; Yang, F.; Ji, W. High-mobility transport anisotropy and linear dichroism in few-layer black phosphorus. *Nat. Commun.* **2014**, *5*, 4475. [CrossRef] [PubMed]
36. Tran, V.; Soklaski, R.; Liang, Y.; Yang, L. Layer-controlled band gap and anisotropic excitons in few-layer black phosphorus. *Phys. Rev. B* **2014**, *89*, 235319. [CrossRef]
37. Koenig, S.P.; Doganov, R.A.; Schmidt, H.; Castro Neto, A.H.; Özyilmaz, B. Electric field effect in ultrathin black phosphorus. *Appl. Phys. Lett.* **2014**, *104*, 103106. [CrossRef]
38. Andres, C.-G.; Leonardo, V.; Elsa, P.; Joshua, O.I.; Narasimha-Acharya, K.L.; Sofya, I.B.; Dirk, J.G.; Michele, B.; Gary, A.S.; Alvarez, J.V.; et al. Isolation and characterization of few-layer black phosphorus. *2D Mater.* **2014**, *1*, 025001.

39. Buscema, M.; Groenendijk, D.J.; Blanter, S.I.; Steele, G.A.; van der Zant, H.S.J.; Castellanos-Gomez, A. Fast and broadband photoresponse of few-layer black phosphorus field-effect transistors. *Nano Lett.* **2014**, *14*, 3347–3352. [CrossRef] [PubMed]
40. Mannix, A.J.; Zhou, X.-F.; Kiraly, B.; Wood, J.D.; Alducin, D.; Myers, B.D.; Liu, X.; Fisher, B.L.; Santiago, U.; Guest, J.R.; et al. Synthesis of borophenes: Anisotropic, two-dimensional boron polymorphs. *Science* **2015**, *350*, 1513–1516. [CrossRef] [PubMed]
41. Osada, M.; Sasaki, T. Two-dimensional dielectric nanosheets: Novel nanoelectronics from nanocrystal building blocks. *Adv. Mater.* **2012**, *24*, 210–228. [CrossRef] [PubMed]
42. Nair, R.R.; Blake, P.; Grigorenko, A.N.; Novoselov, K.S.; Booth, T.J.; Stauber, T.; Peres, N.M.R.; Geim, A.K. Fine structure constant defines visual transparency of graphene. *Science* **2008**, *320*, 1308. [CrossRef] [PubMed]
43. Eda, G.; Maier, S.A. Two-dimensional crystals: Managing light for optoelectronics. *ACS Nano* **2013**, *7*, 5660–5665. [CrossRef] [PubMed]
44. Geim, A.K.; Grigorieva, I.V. Van der waals heterostructures. *Nature* **2013**, *499*, 419–425. [CrossRef] [PubMed]
45. Koenderink, A.F.; Alù, A.; Polman, A. Nanophotonics: Shrinking light-based technology. *Science* **2015**, *348*, 516–521. [CrossRef] [PubMed]
46. Bao, Q.; Loh, K.P. Graphene photonics, plasmonics, and broadband optoelectronic devices. *ACS Nano* **2012**, *6*, 3677–3694. [CrossRef] [PubMed]
47. Grigorenko, A.N.; Polini, M.; Novoselov, K.S. Graphene plasmonics. *Nat. Photonics* **2012**, *6*, 749–758. [CrossRef]
48. Brar, V.W.; Jang, M.S.; Sherrott, M.; Lopez, J.J.; Atwater, H.A. Highly confined tunable mid-infrared plasmonics in graphene nanoresonators. *Nano Lett.* **2013**, *13*, 2541–2547. [CrossRef] [PubMed]
49. Brar, V.W.; Jang, M.S.; Sherrott, M.; Kim, S.; Lopez, J.J.; Kim, L.B.; Choi, M.; Atwater, H. Hybrid surface-phonon-plasmon polariton modes in graphene/monolayer h-bn heterostructures. *Nano Lett.* **2014**, *14*, 3876–3880. [CrossRef] [PubMed]
50. Jang, M.S.; Brar, V.W.; Sherrott, M.C.; Lopez, J.J.; Kim, L.; Kim, S.; Choi, M.; Atwater, H.A. Tunable large resonant absorption in a midinfrared graphene salisbury screen. *Phys. Rev. B* **2014**, *90*, 165409. [CrossRef]
51. Brar, V.W.; Sherrott, M.C.; Jang, M.S.; Kim, S.; Kim, L.; Choi, M.; Sweatlock, L.A.; Atwater, H.A. Electronic modulation of infrared radiation in graphene plasmonic resonators. *Nat. Commun.* **2015**, *6*, 7032. [CrossRef] [PubMed]
52. Jariwala, D.; Davoyan, A.R.; Tagliabue, G.; Sherrott, M.C.; Wong, J.; Atwater, H.A. Near-unity absorption in van der waals semiconductors for ultrathin optoelectronics. *Nano Lett.* **2016**, *16*, 5482–5487. [CrossRef] [PubMed]
53. Xia, F.; Wang, H.; Xiao, D.; Dubey, M.; Ramasubramaniam, A. Two-dimensional material nanophotonics. *Nat. Photonics* **2014**, *8*, 899–907. [CrossRef]
54. Buscema, M.; Island, J.O.; Groenendijk, D.J.; Blanter, S.I.; Steele, G.A.; van der Zant, H.S.J.; Castellanos-Gomez, A. Photocurrent generation with two-dimensional van der waals semiconductors. *Chem. Soc. Rev.* **2015**, *44*, 3691–3718. [CrossRef] [PubMed]
55. Bhimanapati, G.R.; Lin, Z.; Meunier, V.; Jung, Y.; Cha, J.; Das, S.; Xiao, D.; Son, Y.; Strano, M.S.; Cooper, V.R.; et al. Recent advances in two-dimensional materials beyond graphene. *ACS Nano* **2015**, *9*, 11509–11539. [CrossRef] [PubMed]
56. Butler, S.Z.; Hollen, S.M.; Cao, L.; Cui, Y.; Gupta, J.A.; Gutiérrez, H.R.; Heinz, T.F.; Hong, S.S.; Huang, J.; Ismach, A.F.; et al. Progress, challenges, and opportunities in two-dimensional materials beyond graphene. *ACS Nano* **2013**, *7*, 2898–2926. [CrossRef] [PubMed]
57. Xu, M.; Liang, T.; Shi, M.; Chen, H. Graphene-like two-dimensional materials. *Chem. Rev.* **2013**, *113*, 3766–3798. [CrossRef] [PubMed]
58. Howell, S.L.; Jariwala, D.; Wu, C.-C.; Chen, K.-S.; Sangwan, V.K.; Kang, J.; Marks, T.J.; Hersam, M.C.; Lauhon, L.J. Investigation of band-offsets at monolayer–multilayer MoS_2 junctions by scanning photocurrent microscopy. *Nano Lett.* **2015**, *15*, 2278–2284. [CrossRef] [PubMed]
59. Wu, C.-C.; Jariwala, D.; Sangwan, V.K.; Marks, T.J.; Hersam, M.C.; Lauhon, L.J. Elucidating the photoresponse of ultrathin MoS_2 field-effect transistors by scanning photocurrent microscopy. *J. Phys. Chem. Lett.* **2013**, *4*, 2508–2513. [CrossRef]
60. Furchi, M.M.; Polyushkin, D.K.; Pospischil, A.; Mueller, T. Mechanisms of photoconductivity in atomically thin MoS_2. *Nano Lett.* **2014**, *14*, 6165–6170. [CrossRef] [PubMed]

61. Park, J.; Ahn, Y.H.; Ruiz-Vargas, C. Imaging of photocurrent generation and collection in single-layer graphene. *Nano Lett.* **2009**, *9*, 1742–1746. [CrossRef] [PubMed]

62. Xia, F.; Mueller, T.; Golizadeh-Mojarad, R.; Freitag, M.; Lin, Y.-M.; Tsang, J.; Perebeinos, V.; Avouris, P. Photocurrent imaging and efficient photon detection in a graphene transistor. *Nano Lett.* **2009**, *9*, 1039–1044. [CrossRef] [PubMed]

63. Yin, Z.; Li, H.; Li, H.; Jiang, L.; Shi, Y.; Sun, Y.; Lu, G.; Zhang, Q.; Chen, X.; Zhang, H. Single-layer MoS$_2$ phototransistors. *ACS Nano* **2012**, *6*, 74–80. [CrossRef] [PubMed]

64. Nolas, G.S.; Sharp, J.; Goldsmid, J. *Thermoelectrics: Basic Principles and New Materials Developments*; Springer Science & Business Media: New York, NY, USA, 2013; Volume 45.

65. Cutler, M.; Mott, N.F. Observation of anderson localization in an electron gas. *Phys. Rev.* **1969**, *181*, 1336–1340. [CrossRef]

66. Xu, X.; Gabor, N.M.; Alden, J.S.; van der Zande, A.M.; McEuen, P.L. Photo-thermoelectric effect at a graphene interface junction. *Nano Lett.* **2010**, *10*, 562–566. [CrossRef] [PubMed]

67. Seol, J.H.; Jo, I.; Moore, A.L.; Lindsay, L.; Aitken, Z.H.; Pettes, M.T.; Li, X.; Yao, Z.; Huang, R.; Broido, D.; et al. Two-dimensional phonon transport in supported graphene. *Science* **2010**, *328*, 213–216. [CrossRef] [PubMed]

68. Grosse, K.L.; Bae, M.-H.; Lian, F.; Pop, E.; King, W.P. Nanoscale joule heating, peltier cooling and current crowding at graphene-metal contacts. *Nat. Nano* **2011**, *6*, 287–290. [CrossRef] [PubMed]

69. Wang, Z.; Xie, R.; Bui, C.T.; Liu, D.; Ni, X.; Li, B.; Thong, J.T.L. Thermal transport in suspended and supported few-layer graphene. *Nano Lett.* **2011**, *11*, 113–118. [CrossRef] [PubMed]

70. Wu, J.; Schmidt, H.; Amara, K.K.; Xu, X.; Eda, G.; Özyilmaz, B. Large thermoelectricity via variable range hopping in chemical vapor deposition grown single-layer MoS$_2$. *Nano Lett.* **2014**, *14*, 2730–2734. [CrossRef] [PubMed]

71. Buscema, M.; Barkelid, M.; Zwiller, V.; van der Zant, H.S.J.; Steele, G.A.; Castellanos-Gomez, A. Large and tunable photothermoelectric effect in single-layer MoS$_2$. *Nano Lett.* **2013**, *13*, 358–363. [CrossRef] [PubMed]

72. Low, T.; Engel, M.; Steiner, M.; Avouris, P. Origin of photoresponse in black phosphorus phototransistors. *Phys. Rev. B* **2014**, *90*, 081408. [CrossRef]

73. Flores, E.; Ares, J.R.; Castellanos-Gomez, A.; Barawi, M.; Ferrer, I.J.; Sánchez, C. Thermoelectric power of bulk black-phosphorus. *Appl. Phys. Lett.* **2015**, *106*, 022102. [CrossRef]

74. Engel, M.; Steiner, M.; Avouris, P. Black phosphorus photodetector for multispectral, high-resolution imaging. *Nano Lett.* **2014**, *14*, 6414–6417. [CrossRef] [PubMed]

75. Tsen, A.W.; DonevLuke, A.K.; Kurt, H.; Herman, L.H.; Park, J. Imaging the electrical conductance of individual carbon nanotubes with photothermal current microscopy. *Nat. Nano* **2009**, *4*, 108–113. [CrossRef] [PubMed]

76. Gabor, N.M.; Song, J.C.W.; Ma, Q.; Nair, N.L.; Taychatanapat, T.; Watanabe, K.; Taniguchi, T.; Levitov, L.S.; Jarillo-Herrero, P. Hot carrier–assisted intrinsic photoresponse in graphene. *Science* **2011**, *334*, 648–652. [CrossRef] [PubMed]

77. Freitag, M.; Low, T.; Xia, F.; Avouris, P. Photoconductivity of biased graphene. *Nat. Photonics* **2013**, *7*, 53–59. [CrossRef]

78. Hong, T.; Chamlagain, B.; Hu, S.; Weiss, S.M.; Zhou, Z.; Xu, Y.-Q. Plasmonic hot electron induced photocurrent response at MoS$_2$–metal junctions. *ACS Nano* **2015**, *9*, 5357–5363. [CrossRef] [PubMed]

79. Wang, T.; Hu, S.; Chamlagain, B.; Hong, T.; Zhou, Z.; Weiss, S.M.; Xu, Y.-Q. Visualizing light scattering in silicon waveguides with black-phosphorus photodetectors. *Adv. Mater.* **2016**, *28*, 7162–7166. [CrossRef] [PubMed]

80. Néstor, P.-L.; Zhong, L.; Nihar, R.P.; Agustín, I.-R.; Ana Laura, E.; Amber, M.; Jun, L.; Pulickel, M.A.; Humberto, T.; Luis, B.; et al. Cvd-grown monolayered MoS$_2$ as an effective photosensor operating at low-voltage. *2D Mater.* **2014**, *1*, 011004.

81. Mueller, T.; Xia, F.; Avouris, P. Graphene photodetectors for high-speed optical communications. *Nat. Photonics* **2010**, *4*, 297–301. [CrossRef]

82. Zhang, W.; Chiu, M.-H.; Chen, C.-H.; Chen, W.; Li, L.-J.; Wee, A.T.S. Role of metal contacts in high-performance phototransistors based on WSe$_2$ monolayers. *ACS Nano* **2014**, *8*, 8653–8661. [CrossRef] [PubMed]

83. Hong, T.; Chamlagain, B.; Lin, W.; Chuang, H.-J.; Pan, M.; Zhou, Z.; Xu, Y.-Q. Polarized photocurrent response in black phosphorus field-effect transistors. *Nanoscale* **2014**, *6*, 8978–8983. [CrossRef] [PubMed]

84. Wang, H.; Liu, F.; Fu, W.; Fang, Z.; Zhou, W.; Liu, Z. Two-dimensional heterostructures: Fabrication, characterization, and application. *Nanoscale* **2014**, *6*, 12250–12272. [CrossRef] [PubMed]
85. Withers, F.; Del Pozo-Zamudio, O.; Mishchenko, A.; Rooney, A.P.; Gholinia, A.; Watanabe, K.; Taniguchi, T.; Haigh, S.J.; Geim, A.K.; Tartakovskii, A.I.; et al. Light-emitting diodes by band-structure engineering in van der waals heterostructures. *Nat. Mater.* **2015**, *14*, 301–306. [CrossRef] [PubMed]
86. Niu, T.; Li, A. From two-dimensional materials to heterostructures. *Prog. Surface Sci.* **2015**, *90*, 21–45. [CrossRef]
87. Jariwala, D.; Howell, S.L.; Chen, K.-S.; Kang, J.; Sangwan, V.K.; Filippone, S.A.; Turrisi, R.; Marks, T.J.; Lauhon, L.J.; Hersam, M.C. Hybrid, gate-tunable, van der waals p–n heterojunctions from pentacene and MoS₂. *Nano Lett.* **2016**, *16*, 497–503. [CrossRef] [PubMed]
88. Jariwala, D.; Sangwan, V.K.; Wu, C.-C.; Prabhumirashi, P.L.; Geier, M.L.; Marks, T.J.; Lauhon, L.J.; Hersam, M.C. Gate-tunable carbon nanotube–MoS₂ heterojunction p-n diode. *Proc. Natl. Acad. Sci. USA* **2013**, *110*, 18076–18080. [CrossRef] [PubMed]
89. Jariwala, D.; Marks, T.J.; Hersam, M.C. Mixed-dimensional van der waals heterostructures. *Nat. Mater.* **2016**. [CrossRef] [PubMed]
90. Deng, Y.; Luo, Z.; Conrad, N.J.; Liu, H.; Gong, Y.; Najmaei, S.; Ajayan, P.M.; Lou, J.; Xu, X.; Ye, P.D. Black phosphorus–monolayer MoS₂ van der waals heterojunction p–n diode. *ACS Nano* **2014**, *8*, 8292–8299. [CrossRef] [PubMed]
91. Hong, T.; Chamlagain, B.; Wang, T.; Chuang, H.-J.; Zhou, Z.; Xu, Y.-Q. Anisotropic photocurrent response at black phosphorus-MoS₂ p-n heterojunctions. *Nanoscale* **2015**, *7*, 18537–18541. [CrossRef] [PubMed]
92. Wang, L.; Meric, I.; Huang, P.Y.; Gao, Q.; Gao, Y.; Tran, H.; Taniguchi, T.; Watanabe, K.; Campos, L.M.; Muller, D.A.; et al. One-dimensional electrical contact to a two-dimensional material. *Science* **2013**, *342*, 614–617. [CrossRef] [PubMed]
93. Liu, W.; Lee, B.; Naylor, C.H.; Ee, H.-S.; Park, J.; Johnson, A.T.C.; Agarwal, R. Strong exciton–plasmon coupling in MoS₂ coupled with plasmonic lattice. *Nano Lett.* **2016**, *16*, 1262–1269. [CrossRef] [PubMed]
94. Lee, B.; Park, J.; Han, G.H.; Ee, H.-S.; Naylor, C.H.; Liu, W.; Johnson, A.T.C.; Agarwal, R. Fano resonance and spectrally modified photoluminescence enhancement in monolayer MoS₂ integrated with plasmonic nanoantenna array. *Nano Lett.* **2015**, *15*, 3646–3653. [CrossRef] [PubMed]
95. Sobhani, A.; Lauchner, A.; Najmaei, S.; Ayala-Orozco, C.; Wen, F.; Lou, J.; Halas, N.J. Enhancing the photocurrent and photoluminescence of single crystal monolayer MoS₂ with resonant plasmonic nanoshells. *Appl. Phys. Lett.* **2014**, *104*, 031112. [CrossRef]
96. Butun, S.; Tongay, S.; Aydin, K. Enhanced light emission from large-area monolayer MoS₂ using plasmonic nanodisc arrays. *Nano Lett.* **2015**, *15*, 2700–2704. [CrossRef] [PubMed]
97. Kim, S.; Jang, M.S.; Brar, V.W.; Tolstova, Y.; Mauser, K.W.; Atwater, H.A. Electronically tunable extraordinary optical transmission in graphene plasmonic ribbons coupled to subwavelength metallic slit arrays. *Nat. Commun.* **2016**, *7*, 12323. [CrossRef] [PubMed]
98. Najmaei, S.; Mlayah, A.; Arbouet, A.; Girard, C.; Léotin, J.; Lou, J. Plasmonic pumping of excitonic photoluminescence in hybrid MoS₂–au nanostructures. *ACS Nano* **2014**, *8*, 12682–12689. [CrossRef] [PubMed]
99. Miao, J.; Hu, W.; Jing, Y.; Luo, W.; Liao, L.; Pan, A.; Wu, S.; Cheng, J.; Chen, X.; Lu, W. Surface plasmon-enhanced photodetection in few layer MoS₂ phototransistors with au nanostructure arrays. *Small* **2015**, *11*, 2392–2398. [CrossRef] [PubMed]
100. Wang, W.; Klots, A.; Prasai, D.; Yang, Y.; Bolotin, K.I.; Valentine, J. Hot electron-based near-infrared photodetection using bilayer MoS₂. *Nano Lett.* **2015**, *15*, 7440–7444. [CrossRef] [PubMed]
101. Liu, Y.; Cheng, R.; Liao, L.; Zhou, H.; Bai, J.; Liu, G.; Liu, L.; Huang, Y.; Duan, X. Plasmon resonance enhanced multicolour photodetection by graphene. *Nat. Commun.* **2011**, *2*, 579. [CrossRef] [PubMed]
102. Echtermeyer, T.J.; Britnell, L.; Jasnos, P.K.; Lombardo, A.; Gorbachev, R.V.; Grigorenko, A.N.; Geim, A.K.; Ferrari, A.C.; Novoselov, K.S. Strong plasmonic enhancement of photovoltage in graphene. *Nat. Commun.* **2011**, *2*, 458. [CrossRef] [PubMed]
103. Youngblood, N.; Chen, C.; Koester, S.J.; Li, M. Waveguide-integrated black phosphorus photodetector with high responsivity and low dark current. *Nat. Photonics* **2015**, *9*, 247–252. [CrossRef]
104. Gan, X.; Shiue, R.-J.; Gao, Y.; Meric, I.; Heinz, T.F.; Shepard, K.; Hone, J.; Assefa, S.; Englund, D. Chip-integrated ultrafast graphene photodetector with high responsivity. *Nat. Photonics* **2013**, *7*, 883–887. [CrossRef]

105. Pospischil, A.; Humer, M.; Furchi, M.M.; Bachmann, D.; Guider, R.; Fromherz, T.; Mueller, T. Cmos-compatible graphene photodetector covering all optical communication bands. *Nat. Photonics* **2013**, *7*, 892–896. [CrossRef]

106. Wang, X.; Cheng, Z.; Xu, K.; Tsang, H.K.; Xu, J.-B. High-responsivity graphene/silicon-heterostructure waveguide photodetectors. *Nat. Photonics* **2013**, *7*, 888–891. [CrossRef]

107. Liu, M.; Yin, X.; Ulin-Avila, E.; Geng, B.; Zentgraf, T.; Ju, L.; Wang, F.; Zhang, X. A graphene-based broadband optical modulator. *Nature* **2011**, *474*, 64–67. [CrossRef] [PubMed]

108. Ding, Y.; Zhu, X.; Xiao, S.; Hu, H.; Frandsen, L.H.; Mortensen, N.A.; Yvind, K. Effective electro-optical modulation with high extinction ratio by a graphene-silicon microring resonator. *Nano Lett.* **2015**, *15*, 4393–4400. [CrossRef] [PubMed]

109. Gan, X.; Mak, K.F.; Gao, Y.; You, Y.; Hatami, F.; Hone, J.; Heinz, T.F.; Englund, D. Strong enhancement of light-matter interaction in graphene coupled to a photonic crystal nanocavity. *Nano Lett.* **2012**, *12*, 5626–5631. [CrossRef] [PubMed]

110. Majumdar, A.; Kim, J.; Vuckovic, J.; Wang, F. Electrical control of silicon photonic crystal cavity by graphene. *Nano Lett.* **2013**, *13*, 515–518. [CrossRef] [PubMed]

111. Gan, X.; Shiue, R.-J.; Gao, Y.; Mak, K.F.; Yao, X.; Li, L.; Szep, A.; Walker, D.; Hone, J.; Heinz, T.F.; et al. High-contrast electrooptic modulation of a photonic crystal nanocavity by electrical gating of graphene. *Nano Lett.* **2013**, *13*, 691–696. [CrossRef] [PubMed]

112. Wu, S.; Buckley, S.; Schaibley, J.R.; Feng, L.; Yan, J.; Mandrus, D.G.; Hatami, F.; Yao, W.; Vuckovic, J.; Majumdar, A.; et al. Monolayer semiconductor nanocavity lasers with ultralow thresholds. *Nature* **2015**, *520*, 69–72. [CrossRef] [PubMed]

113. Majumdar, A.; Dodson, C.M.; Fryett, T.K.; Zhan, A.; Buckley, S.; Gerace, D. Hybrid 2D material nanophotonics: A scalable platform for low-power nonlinear and quantum optics. *ACS Photonics* **2015**, *2*, 1160–1166. [CrossRef]

114. Kunz, K.S.; Luebbers, R.J. *The finite difference time domain method for electromagnetics*; CRC press: Boca Raton, FL, USA, 1993.

115. Hu, S.; Weiss, S.M. Design of photonic crystal cavities for extreme light concentration. *ACS Photonics* **2016**, *3*, 1647–1653. [CrossRef]

116. Ee, H.-S.; Kang, J.-H.; Brongersma, M.L.; Seo, M.-K. Shape-dependent light scattering properties of subwavelength silicon nanoblocks. *Nano Lett.* **2015**, *15*, 1759–1765. [CrossRef] [PubMed]

117. Dunn, R.C. Near-field scanning optical microscopy. *Chem. Rev.* **1999**, *99*, 2891–2928. [CrossRef] [PubMed]

electronics

MDPI

Review

Recent Advances in Electronic and Optoelectronic Devices Based on Two-Dimensional Transition Metal Dichalcogenides

Mingxiao Ye, Dongyan Zhang and Yoke Khin Yap *

Department of Physics, Michigan Technological University, 1400 Townsend Drive, Houghton, MI 49931, USA; mye1@mtu.edu (M.Y.); dozhang@mtu.edu (D.Z.)
* Correspondence: ykyap@mtu.edu; Tel.: +1-906-487-2900

Academic Editor: Zhenqiang Ma
Received: 16 March 2017; Accepted: 12 May 2017; Published: 2 June 2017

Abstract: Two-dimensional transition metal dichalcogenides (2D TMDCs) offer several attractive features for use in next-generation electronic and optoelectronic devices. Device applications of TMDCs have gained much research interest, and significant advancement has been recorded. In this review, the overall research advancement in electronic and optoelectronic devices based on TMDCs are summarized and discussed. In particular, we focus on evaluating field effect transistors (FETs), photovoltaic cells, light-emitting diodes (LEDs), photodetectors, lasers, and integrated circuits (ICs) using TMDCs.

Keywords: two-dimensional materials; transition metal dichalcogenides; heterostructures; heterojunctions; field effect transistors; photovoltaic cells; light-emitting diodes; photodetectors; lasers; integrated circuits; MoS_2; WS_2; $MoSe_2$; WSe_2; $MoTe_2$; ReS_2; $ReSe_2$

1. Introduction

The miniaturization of silicon-based transistors has led to unprecedented progress in smaller and faster electronic devices including smartphones and tablet computers. Further miniaturization of transistors is hindered by some fundamental limits including high contact resistance, short channel effects, and high leakage currents [1–3]. Tremendous efforts have been attempted to overcoming these limitations using novel device architecture [4,5]. In addition, silicon-based devices are not applicable for flexible electronics. All the shortcomings of silicon-based transistors have led to the motivation for the search of new device concepts and alternative materials. For example, "transistors without using semiconductors" have been proposed using boron nitride nanotubes functionalized with gold quantum-dots (QDs-BNNTs) [6,7] and graphene–BNNT heterojunctions [8]. All these materials are applicable for flexible electronics. On the other hand, two-dimensional transition metal dichalcogenides (2D TMDCs) offer several attractive features for next-generation transistors, including high electron mobility (up to ~1000 cm^2 $V^{-1} \cdot s^{-1}$ for MoS_2), atomically thin and flexible. In addition, TMDCs can be free of dangling bonds on their surfaces, and offer many interesting optical properties for their potential application in optoelectronic devices.

Although research publication on TMDCs was started around the 1960s, publication activity has increased significantly since 2014. According to our analysis in Web of Science^{TM}, a total of 3829 journal articles were published up to 2016, with 33.8%, 21.5%, 12.5%, and 5.5% of these articles were published in 2016, 2015, 2014, and 2013, respectively. Based on these recent publications, we will discuss the application of 2D TMDCs in field effect transistors (FETs), optoelectronic devices (photovoltaic cells, light-emitting diodes, and lasers), and integrated circuits (ICs). Interested readers may refer to a recent review on the same topics [9]. For the convenience of the readers, we summarize a series of TMDCs along with the nature of their band gaps in Table 1.

Table 1. Summary of semiconductor TMDCs materials (MX$_2$) and the nature of their band gaps for monolayer (1L) to bulk samples. The direct (D) and indirect (I) bandgaps are as denoted. These values are extracted from both theoretical (T) and experimental (E) study. All data are referring to optical bandgaps at room temperature without considering quasiparticle (e.g., excitons) effects.

M or X	S	Se	Te
Mo	1L: ~1.8–1.9 eV (D) [10] Bulk: 1.2 eV (I) [10]	1L: 1.34 eV (D)(T) [11] 1L: 1.58 eV (D)(E) [12] Bulk: 1.1 eV (I)(E) [13] Bulk: 1.1 eV (I)(T) [14]	1L: 1.07 eV (D)(T) [15] 1L: 1.1 eV (D)(E) [16] Bulk: 1.0 eV (I)(E) [17] 0.82 eV (I)(T) [18]
W	1L: 1.94 eV (D)(T) [19] 1L: 2.14 eV (D)(E) [20] Bulk: 1.35 eV (I)(E) [21]	1L: 1.74 eV (D)(T) [19] 1L: 1.65 eV (D)(E) [22] Bulk: 1.1 eV (I)(D) [23] Bulk: 1.2 eV (I)(E) [21]	1L: 1.14 eV (D)(T) [19] Bulk: 0.7 eV (I)(T) [23]
Re	1L: 1.43 eV (D)(T) [24] 1L: 1.55 eV (D)(E) [24] Bulk: 1.35 eV (D)(T) [24] Bulk: 1.47 eV (D)(E) [25]	1L: 1.34 eV (I)(T) [26] 1L: 1.239 eV (D)(T) [27] 1L: 1.47 eV (I)(E) [28] 2L: 1.165 eV (D)(T) [27] 2L: 1.32 eV (I)(E) [29] 4L: 1.092 eV (D)(T) [27] Bulk: 1.06 eV (I)(T) [26] Bulk: 1.18 (I)(E) [30]	

2. Field Effect Transistors

Field-effect transistors (FETs) are among the basic devices for modern electronic circuits. A typical FET modulates the conductivity of the semiconductor channel between the source (S) and the drain (D) terminals by controlling the applied voltage on the gate (G) electrode. Some of the important figures of merit for FETs are defined here. First, a good FET should have a low subthreshold swing/slope:

$$SS = \left(\frac{d(\log I_{DS})}{dV_G} \right)^{-1}$$

which is defined as the gate voltage (V_G) required to change the source–drain current (I_{DS}) by one order of magnitude (in the unit of volt per decade). FETs are treated as switches that offer a high conductivity of the channel (ON state), and vice versa (OFF state). Hence, the on/off ratio is another crucial factor in measuring the performance of FETs. In addition, mobility is important for high-performance FETs. In 2D TMDCs, the transport of charge carriers is confined to the plane of the materials. The mobility of carriers is related to the scattering by

$$\mu = \frac{e\tau}{m^*}$$

where τ is the scattering time, and m^* is the effective mass of in-plane electron. There are following scattering mechanisms: (1) electron–phonon scattering, including longitudinal acoustic (LA), transverse acoustic (TA) [31,32], polar (longitudinal) optical (LO), and homopolar optical phonons [33]; (2) electron–electron (Coulomb) scattering [34] and charged impurities scattering [35]; (3) surface (interface) roughness Coulomb scattering [34] and phonon scattering [36]; and (4) short-range scattering, which is from the defect and dislocation of the lattice [35,37,38]. The degree to which these mechanics affect the mobility can be evaluated by Matthiessen's rule:

$$\frac{1}{\mu} = \frac{1}{\mu_1} + \frac{1}{\mu_2} + \frac{1}{\mu_3} + \frac{1}{\mu_4}.$$

Finally, Flicker noise is a factor that limits the performance of the FETs. This is a type of electronic noise with a frequency spectrum such that the power density spectrum is inversely proportional to frequency, f, and is therefore referred as 1/f noise. Flicker noise occurs in almost all electronic devices and often shows up as the resistance fluctuation. It is generally related to the impurities in the conductive channel and generation/recombination noise in transistors due to memory of the base current [39].

2.1. Transistors with Multilayered TMDCs

Although the electrical properties and electronic structures of TMDCs were studied in the 1960s [33], their device applications have been limited until 2004. One of the earliest works in which TMDCs were used for FETs was reported by comparing WSe_2 with single-crystal Si metal oxide FETs. A p-type conductivity as high as 500 $cm^2 \cdot V^{-1} \cdot s^{-1}$ was demonstrated at room temperature with a 10^4 on/off ratio at 60 K [40]. Thus, a further study based on multilayer MoS_2 FETs with back-gate configuration was conducted, but a rather low mobility (0.1–40 $cm^2 \cdot V^{-1} \cdot s^{-1}$) was demonstrated [41–43]. Apparently, the reported mobility from multilayered MoS_2 was quite lower as compared to those reported for bulk MoS_2, where mobility is often ranged from 100 to 260 $cm^2 \cdot V^{-1} \cdot s^{-1}$ [33]. Table 2 summarizes the performances of FETs constructed by various TMDCs. It is noted that the mobility of graphene was much higher, up to 10,000 $cm^2 \cdot V^{-1} \cdot s^{-1}$, even when using SiO_2 as the gate materials [44].

Table 2. Performance of FETs based on multilayer TMDCs prepared by mechanical exfoliation and by chemical vapor deposition (as noted with CVD). Electron mobility is compiled here (or denoted by e). The hole mobility is denoted by h.

Materials	Configuration (Method)	Mobility ($cm^2 \cdot V^{-1} \cdot s^{-1}$)	On/Off Ratio	Subthreshold Swing ($mV \cdot dec^{-1}$)	Temperature (K)	Reference
MoS_2	Back-gated	3			300	[41]
	Back-gated	40	10^5	1000	300	[42]
	Back-gated (CVD)	2×10^{-2}	10^4		300	[45]
	Back-gated	100	10^6	80	300	[46]
	Back-gated	2.4	10^7		300	[47]
	Dual-gated	517	10^8	140	300	[48]
	Back-gated	700			300	[49]
	Back-gated (CVD)	17	4×10^8(bi-) 10^4(multi-)		300	[50]
	Four-terminal Back-gated	306.5	10^6		300	[51]
	Four-terminal Top-gated	470(e)/480(h)			300	[52]
	ZrO_2 & CNT Back-gated		10^6		3	[53]
$MoSe_2$	Back-gated	50	10^6		300	[54]
	Four-terminal Back-gated on SiO_2/parylene-C (CVD)	50(Si) 160(parylene-C) 500	10^6(e) 10^3(h)		295/100	[55]
	Four-terminal Back-gated (CVD)	200(e)/150(h)	10^6		275	[56]
	Back-gated (CVD)	10	10^3		300	[57]

Table 2. *Cont.*

Materials	Configuration (Method)	Mobility $(cm^2 \cdot V^{-1} \cdot s^{-1})$	On/Off Ratio	Subthreshold Swing $(mV \cdot dec^{-1})$	Temperature (K)	Reference
	Back-gated	0.03(e)/0.3(h)	2×10^3		300	[58]
	Four-terminal Back-gated	20(h)	10^5	140	300	[59]
	Ionic Liquid Top-gated	30(e)/10(h)		140(e)/125(h)	300	[60]
$MoTe_2$	Back-gated	6(h)	10^5		300	[61]
	Solid Polymer Electrolyte Back-gated	7(e)/26(h)	10^5	90	300	[62]
	Back-gated	25.2(e)/1.5(h)	2.1×10^5(e)/ 5.7×10^4(h)		280	[63]
	Back-gated	2.04(h)			300	[64]
	Iodine-transport Back-gated (PVD)		10^5		300	[65]
	Ionic Liquid Top-gated	20(e)/90(h)		90	300	[66]
WS_2	Ionic Liquid Top-gated	19(e)/12(h)	10^6	63(e)/67(h)	300	[20]
	Back-gated	234	10^8		300	[67]
	Four-terminal Back-gated	20	10^6	70	300	[68]
	Four-terminal Top-gated	500(h)			300	[40]
	Ionic Liquid Top-gated	200(e/h) 330(e)/270(h)	10		170/160/77	[69]
	Back-gated (CVD)	350(h)	10^8		300	[70]
WSe_2	Back-gated (CVD)	650(h)	10^6	250/140	150/300/105	[71]
	Back-gated (CVD)	10(h)	10^4		300	[72]
	Back-gated	92(h)	~10		300	[73]
	Dual-gated	12/26	10^5	148	300/77	[74]
	Dual-gated	1/5	10^6	750	300/120	[75]
	Back-gated	15.4	10^7	100	300	[76]
ReS_2	Back-gated	1.5	10^5		300	[77]
	Back-gated	11	3×10^5		300	[78]
	Back-gated (CVD)	7.2×10^{-2}	10^3		300	[79]
	Top-gated	0.1			300	[27]
$ReSe_2$	Back-gated	6.7	10^5	1300	300	[80]
	Back-gated (CVD)	1.36×10^{-3}(h)			300	[81]

Later, by using high-k dielectrics (50-nm-thick Al_2O_3), FETs with a back-gate configuration were reported with relatively high mobility (>100 $cm^2 \cdot V^{-1} \cdot s^{-1}$) and on/off ratio (>$10^6$) based on multilayer MoS_2 (Figure 1a). The width and length of the device were 4 μm and 7 μm, respectively. Thirty-nanometer-thick multilayer MoS_2 was deposited on a silicon substrate that was coated with a 50-nm-thick Al_2O_3 layer. As shown in Figure 1b,c, the near-ideal subthreshold slope/swing of 80 mV/dec and robust current saturation over a large voltage window are demonstrated [46]. Accumulation mode occurred at a positive gate bias, and the window of depletion mode occurred at negative gate bias. At large negative gate biases, the drain current recovers and forms an inversion mode. Furthermore, FETs with Schottky contacts and Al_2O_3 gate oxide (15-nm-thick) on the top of MoS_2 channel with back-gated configuration was shown to have mobility as high as 700 cm$^2 \cdot$V$^{-1} \cdot$s^{-1} at room temperature [49]. Liu et al. have further demonstrated that the field effect mobility of multilayer (20 layers) MoS_2 FETs exceeds 500 cm$^2 \cdot$V$^{-1} \cdot$s^{-1} due to the smaller bandgap (thus a smaller Schottky barrier) [48], compared with monolayer MoS_2 FET within the same top-gated configuration [82].

The dependence of carrier mobility on temperature in multilayer MoS_2-based FETs is due to different scattering mechanisms as shown in Figure 2. At low temperature, the carrier mobility is limited by charged impurity scattering. The mobility is determined by the combined effect of the homopolar phonon and the polar-optical phonon scatterings at room temperature [46]. Although the high-κ dielectric materials (e.g., Al_2O_3 and HfO_2) may screen Coulomb scattering from charged impurities [49], the complete recovery of the intrinsic phonon-limited mobility has not been observed in high-κ dielectrics encapsulated TMDCs devices at room temperature.

Another limiting factor of mobility is the substrates. It was shown that trapped charges in the interface of MoS_2–SiO_2 would lead to carrier localization in the bilayer and trilayer MoS_2 channel [83]. Considerable mobility improvement was reported in multilayer MoS_2 on PMMA dielectric in comparison with MoS_2 on SiO_2, which was attributed to the reduced short-range disorder or long-range disorder at the MoS_2/PMMA interface [52]. Other phenomena, such as thickness-independent SS and hysteresis as the gate swept, have been attributed to the interface traps between the bottom layer of WS_2/$MoTe_2$ and SiO_2 substrate [64,68]. Besides, the presence of a low-energy optical phonon mode in SiO_2 (~60 meV) may also cause non-negligible surface polar optical scattering. A two to three times mobility improvement is consistently observed in $MoSe_2$ FETs on parylene-C substrates compared to SiO_2 [55]. WSe_2 FETs with parylene top-gate dielectric have demonstrated high room-temperature mobility up to 500 cm$^2 \cdot$V$^{-1} \cdot$s^{-1} [40].

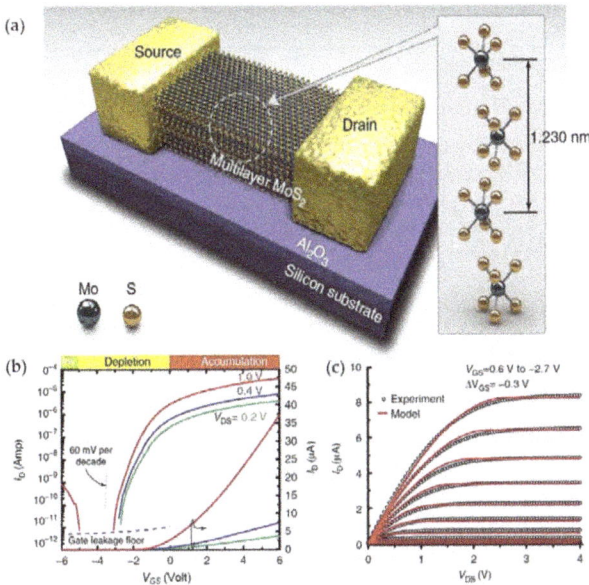

Figure 1. (a) A MoS_2 thin film transistor and its transport properties. (b) Drain current (I_D) versus back-gate bias (V_{GS}) with different drain bias (V_{DS}). (c) Drain current (I_D) versus drain bias (V_{DS}) showing current saturation, including a long-channel model (red lines) showing excellent agreement with measured device behavior. Reproduced with permission from [46], Nature Publishing Group, 2012.

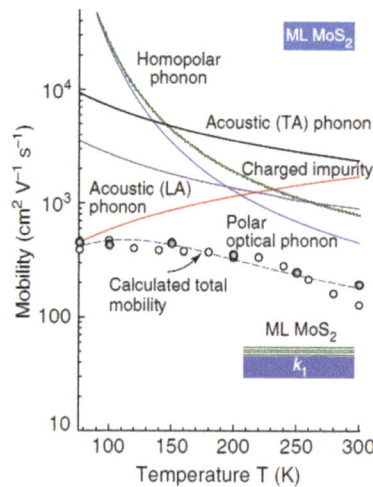

Figure 2. Charge transport properties of the multilayer MoS$_2$ channel. Measured temperature-dependent mobility of MoS$_2$ field effect transistors (FETs). Reproduced with permission from [46], Nature Publishing Group, 2012.

Low carrier mobility in TMDC devices might be due to the shortcomings of two-terminal measurements. A mobility of 306.5 cm$^2 \cdot$V$^{-1} \cdot$s^{-1} was extracted from the same multilayer MoS$_2$ device when using a four-terminal configuration, which is considerably greater than 91 cm$^2 \cdot$V$^{-1} \cdot$s^{-1} measured by a two-terminal configuration [51]. This results indicated that the intrinsic carrier mobility on the SiO$_2$ substrate is significantly greater than the values previously reported in MoS$_2$ FETs [42]. The analogous configuration also applied to FETs based on MoS$_2$ [52], MoSe$_2$ [55], MoTe$_2$ [59], WS$_2$ [68], and WSe$_2$ [40]. Therefore, future studies of TMDCs devices should use the four-terminal measurement.

Table 2 summarizes the performances of FETs constructed by various multilayered TMDCs. By comparing the carrier mobility, the on/off ratio, and the subthreshold swing of these devices, we found that devices constructed by CVD MoS$_2$ [45], ReS$_2$ [79], and ReSe$_2$ [81] exhibited very low mobility, which are two or three orders of magnitudes less than those based on exfoliated TMDCs. The relatively low mobility reported in these earlier literatures is likely due to structural defects, such as grain boundary [84]; these grain boundaries act as short-range scattering centers to suppress the carrier transport [85]. After the CVD synthesis conditions were optimized to reduce the grain boundary, the mobility increased to a level similar to that based on exfoliated TMDCs [50,55,70,71]. It was reported that the threshold voltage for devices with exfoliated MoS$_2$ was higher than that based on CVD MoS$_2$ [86]. The possible reasons could be the density of unintentional doping and the Scotch tape residue-induced charges in the exfoliated samples.

Furthermore, as compared to the low-resistance ohmic contacts, the Schottky contacts between the TMDCs and gates are one of the reasons for low carrier mobility. It is believed that the nature of contact is more important than the properties of the TMDCs [87]. By using a non-contact method based on THz-probe spectroscopy, an intrinsic mobility of MoS$_2$ was found to be approaching 4200 cm$^2 \cdot$V$^{-1} \cdot$s^{-1} at 30 K, which is an order of magnitude greater than any other previously reported values in multilayer TMDCs devices [88].

The intrinsic current distribution in FETs based on multilayer MoS$_2$ has been investigated. Here, the 13-layer MoS$_2$ FETs (8 nm in total thickness) were treated as consecutive resistors in a resistor network model that is based on the Thomas−Fermi charge screening and interlayer coupling effects. The authors map the current distribution among the individual layers of the multilayer 2D systems. Result suggests the existence of a centroid of current distribution or the so-called "HOT-SPOT" in multilayer MoS_2 FETs [89]. Figure 3 shows the current distribution in the multilayer MoS_2 FETs at different gate bias conditions. As shown, the current mainly travels through the upper layer that is closer to the source and the drain electrodes. The "HOT-SPOT" still moved toward the top layers even when the gate bias was increased. Hence, the large effective interlayer resistance and charge screening effect will limit the carrier concentration of multilayer FETs.

Recently, FETs based on multilayer MoS_2 with an intrinsic gain over 30, an intrinsic cut-off frequency up to 42 GHz (F_t), and a maximum oscillation frequency up to 50 GHz (F_{max}) were fabricated based on optimum contacts and device geometry [90]. These performances are better than those reported so far for MoS_2 transistors [91]. Multilayer MoS_2 FETs with a 1 nm physical gate length using a single-walled carbon nanotube (SWCNT) as the gate electrode were demonstrated. These ultrashort devices exhibit excellent switching characteristics with a near-ideal subthreshold swing of 65 mV/dec and an on/off current ratio of 10^6. Simulations show an effective channel length of ~3.9 nm in the off state and ~1 nm in the on state [53].

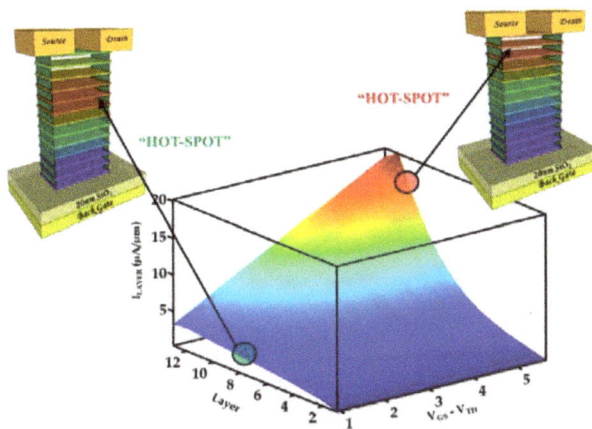

Figure 3. Current distribution among the individual layers of a 13-layer thick MoS_2 FET at different gate bias conditions derived using the resistor network model. The illustrations show the location of the "HOT-SPOT" and the associated current spread schematically corresponding to two different gate bias conditions. The red and blue colors representing the highest and lowest current, respectively. Reproduced with permission from [89], American Chemical Society, 2013.

Finally, while MoS_2 FETs are one of the most studied TMDCs, $MoSe_2$ FETs have been reported to have a low noise characteristic with a Hooge parameter of 3.75×10^{-3} [92], which is similar or lower than MoS_2 FETs [93,94]. On the other hand, WS_2 was once predicted to offer the best transistor performance (the highest on-state current density and mobility) among all TMDCs [95]. However, only a few WS_2 FETs have been reported to have a performance comparable to that of MoS_2 [62]. In addition, $MoTe_2$ offers a favorable bandgap (Table 1) in the near-infrared range. Therefore, $MoTe_2$ FETs are an appropriate candidate to substitute Si and has great potential for high-performance optoelectronic devices [59,63,64]. Finally, ReS_2 and $ReSe_2$ are TMDCs with a direct bandgap in monolayer, multilayer, and bulk forms, as summarized in Tables 1 and 2. They have recently attracted much interest and have been applied to optoelectronic devices.

2.2. Transistors with Monolayer TMDCs

Table 3 summarizes performances of FETs based on monolayer of TMDCs in various gate configurations. As shown, the electron mobility is highest for devices based on exfoliated MoS_2.

Apparently, other TMDCs offer much lower mobility. In addition, the mobility of CVD MoS$_2$ is also significantly lower.

Table 3. Performance of FETs based on monolayer, exfoliated TMDCs, and CVD TMDCs. Electron mobility is compiled here (or denoted by e). The hole mobility is denoted by h.

Materials	Configuration (Method)	Mobility (cm^2·V^{-1}·s^{-1})	On/Off Ratio	Subthreshold Swing (mV·dec^{-1})	Temperature (K)	Reference
MoS$_2$	Top-gated	217	10^8	74	300	[82]
	Simulation Top-gated	350	10^{10}	60	300	[96]
	Back-gated	12			300	[83]
	Top-gated	320	10^6		300	[97]
	Back-gated (CVD)	0.04			300	[98]
	Back-gated	10	10^6		300	[99]
	Polymer Electrolyte Top-gated	10	10^6	60	300	[100]
	Ferro-electric polymer Top-gated	220	10^5	300	300	[101]
	Top-gated	380	10^6	500	300	[102]
	Top-gated	1090	10^8	178	300	[103]
	Back-gated (CVD)	8	10^7		300	[84]
	Multi-terminal Back-gated	1000	10^6		4	[104]
	Back-gated (CVD)	6	10^5		300	[105]
	Top-gated On SiO$_2$/Si$_3$N$_4$ (CVD)	55(Si$_3$N$_4$) 24(SiO$_2$)	10^7(SiO$_2$)		300	[106]
	Top-gated (CVD)	42.3	10^6		300	[107]
	Dual-gated (CVD)	190	10^8	170	300	[108]
MoSe$_2$	Back-gated (CVD)	0.02(e)/0.01(h)	10^2		300	[109]
	Back-gated (CVD)	50	10^6		300	[110]
	Back-gated (CVD)	23(e)/17(h)	10^5		300	[111]
WS$_2$	Ionic Liquid Top-gated	44(e)/43(h)	10^5	52(e)/57(h)	300	[20]
	Back-gated	50/140	10^6		300/83	[112]
	Back-gated (CVD)	4.1	10^5		300	[113]
WSe$_2$	Back-gated	83/337			300/25	[114]
	Top-gated	250(h)	10^6	60	300	[115]
	Ionic Liquid Top-gated	90(e)/7(h)	10^4(e) 10^5(h)		300	[116]
	Polymer electrolyte Back-gated	30(e)/180(h)			300	[117]
	Back-gated (CVD)	100(h)	10^8		300	[70]

Theoretical simulation was applied to predict the performance limits of transistors based on monolayer MoS$_2$. It is known that optical phonon scattering is the intrinsic scattering mechanism that dominates at room temperature. On the other hand, acoustic scattering is dominated at lower temperature (T < 100 K) [118,119]. Figure 4 shows the temperature dependence of electron mobility in freestanding monolayer MoS$_2$ as predicted by first principle calculations. As reported, the mobility (μ) is expected to follow the $T^{-\gamma}$ temperature (T) dependence with γ = 1.69. It was predicted that the mobility of monolayer MoS$_2$ can reach 410 cm$^2 \cdot$V$^{-1} \cdot$s^{-1} at room temperature [120]. This theoretical limit has become the reference for experimental efforts. Indeed, among the TMDCs, the mobilities of MoTe$_2$ (2526 cm$^2 \cdot$V$^{-1} \cdot$s^{-1}), WS$_2$ (1103 cm$^2 \cdot$V$^{-1} \cdot$s^{-1}), and WSe$_2$ (705 cm$^2 \cdot$V$^{-1} \cdot$s^{-1}) were predicted to be greater than the room-temperature phonon-limited electron mobility in the monolayer [121], as a result of low effective mass. However, as shown in Table 3, the reported experimental mobility to date is much lower than that found via theoretical prediction.

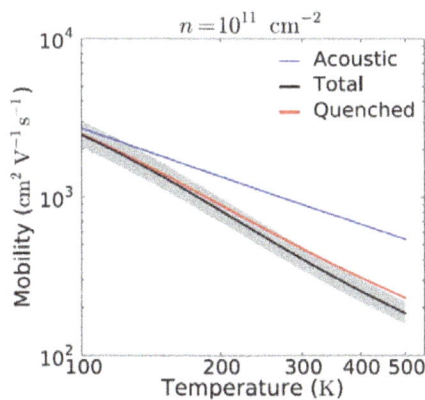

Figure 4. The temperature dependence of the mobility at carrier density $n = 10^{11}$ cm^{-2} calculated with the full collision integral. For comparison, the mobility in the presence of only acoustic deformation potential scattering with the temperature dependence is shown by blue line. The (gray) shaded area shows the variation in the mobility associated with a 10% uncertainty in the calculated deformation potentials. Reproduced with permission from [120], American Physical Society, 2012.

The pioneer work on FETs based on monolayer MoS$_2$ was demonstrated by Radisavljevic et al. in 2011, as shown in Figure 5 [82]. The reported performance includes an excellent on/off current ratio (~10^8), greater mobility (>200 cm$^2 \cdot$V$^{-1} \cdot$s^{-1}), an ideal low subthreshold swing (74 mV/dec), and smaller off-state currents (25 fA/μm). Here, a mechanical exfoliated monolayer MoS$_2$ (0.65 nm) was used as the transport channel and was covered by 30-nm-thick HfO$_2$, which served as the top-gated dielectric layer. The same configuration was also used in FETs with channels made of monolayer WSe$_2$ covered by ZrO$_2$ dielectrics. These FETs also exhibited high room-temperature mobility (~250 cm$^2 \cdot$V$^{-1} \cdot$s^{-1}), a good subthreshold swing of ~60 mV/dec, and a high on/off ratio of 10^6 [115].

It is believed that the encapsulation of 2D-TMDCs by using high dielectric materials (e.g., Al$_2$O$_3$ and HfO$_2$) is critical for achieving high mobility. Authors have suggested that the introduction of high-k dielectrics would strongly dampen the Coulombic scattering in 2D materials due to the dielectric screen [122]. This mobility increase in conjunction with dielectric deposition was also observed in multilayer MoS$_2$ [46] and monolayer MoS$_2$ with polymer electrolytes [100]. The scattering rate is related to the scattering mean free path and the impurity density (i.e., the impurity spacing). Therefore, the charge impurity must be appropriate such that the scattering mean free path is on the same or small order of the phonon mean free path. Therefore, a minimum impurity concentration of 5×10^{11} cm^{-2} is need to dominate phonon scattering [120], which corresponded to the heavy doping by using top-gated dielectric materials. The similar mobility improvement due to such dielectric engineering has been reported earlier for graphene [123–125].

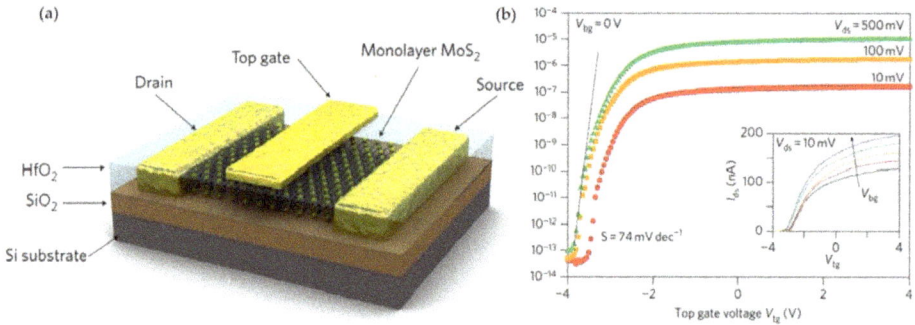

Figure 5. (a) Three-dimensional schematic view of the monolayer MoS_2 transistors. Top gate width and length are 4 um and 500 nm, respectively. (b) I_{ds}–V_{tg} curve recorded for V_{ds} ranging from 10 to 500 mV. Measurements were performed at room temperature with the back-gate grounded. The device can be completely turned off by changing the top gate bias from −2 to −4 V. For V_{ds} = 10 mV, the I_{on}/I_{off} is > 1 × 10^6. For V_{ds} = 500 mV, the I_{on}/I_{off} is > 1 × 10^8 while the subthreshold swing 74 mV/dec. Inset: I_{ds}–V_{tg} for values of V_{bg} = −10, −5, 0, 5, and 10 V. Reproduced with permission from [82], Nature Publishing Group, 2011.

On the other hand, the effects of gate configuration on charge mobility were also studied for FETs based on monolayer MoS_2. Top-gated geometry allows for a lower turn-on voltage, and the integration of multiple devices on the same substrate. Top gate configuration also had the effect of quenching the homopolar phonon mode, which has contributed to a mobility enhancement of 70 $cm^2 \cdot V \cdot s^{-1}$ at room temperature [120]. However, the measured exponent γ factors in the temperature dependence mobility, $\mu = T^{-\gamma}$, are between 0.3 and 0.78 for top-gated devices [126]. These values are much smaller than the theoretically predicted value of 1.52 for monolayer MoS_2 [120] or 2.6 for TMDC bulk crystals [33]. These results suggest that other mechanisms, rather than the quenching of the homopolar phonon, such as the surface roughness scattering created during synthesis, exfoliation, and transferring processes, are affecting the mobility of monolayer MoS_2 in the top-gated devices [127].

In addition, the Schottky barrier height between the electrodes and 2D TMDCs is known to affect carrier mobility of devices. The metals used for electrodes are claimed to define the height of the Schottky barrier associated with their work function. However, Theoretical calculations have shown 2D-TMDCs tend to form high Schottky barriers (0.03 eV to 0.8 eV) with common metals [128–131]. Several experimental practical attempts by inserting 2D layers (Graphene [132,133], hBN [134], Nb doped TMDCs [135], MoO_3 [136]) or thin film (MgO [137], TiO_2 [138]) between TMDCs and metals have been made to reduce Schottky contacts. Besides the handling of TMDC/metal junctions, chemical/electrostatic-doping [69,115] and phase-patterning [139–142] are two successful methods of modifying the electrical properties of channel by reducing contact resistance. It was also reported that in situ annealing at 120 °C for 20 h in vacuum (~10^{-6} mbar) could increase the mobility due to the reduction of Schottky barrier. By this approach, an intrinsic mobility of 1000 $cm^2 \cdot V^{-1} \cdot s^{-1}$ were observed for both monolayer and bilayer devices based on MoS_2 [104].

Theoretical simulation has been investigated for the performance limits of an FET with a 15-nm-long MoS_2 gate length [96]. It has been reported that FETs based on MoS_2 can offer a subthreshold swing as low as 60 mV/dec and a current on/off ratio as high as 10^{10}. The ballistic regime of this device can be as high as 1.6 mA/μm, and the drain-induced barrier lowering (DIBL) is as small as 10 mV/V even with a very short channel length [96]. Thus, TMDCs transistors have the potential to overcome the short-channel effects because of the enhanced gate effect due to their atomic-scale thickness. A similar simulation was reported based on a two-gate MOSFET model with 2D-Silicon and different 2D-TMDCs. Among them, only the monolayer WS$_2$ transistor outperform the 2D silicon and other monolayer TMDC devices in terms of ON-current by about 28.3% [95]. However, these predictions did not consider the effects of edges on the transport properties and therefore require experimental verification.

It should be noted that high transconductance FETs based on monolayer MoS$_2$ with full-channel gates were demonstrated experimentally [103]. As shown in Figure 6a, at low bias V_{ds} = 1 V, the drain–source conductance is close to 0 (g_{ds} < 2 µS/µm, $g_{ds} = \frac{dI_{ds}}{dV_{ds}}$), which means the channel current saturation occur. As shown in Figure 6b, these FETs offer high transconductance $g_m = \frac{dI_{ds}}{dV_{tg}}$ (maximum transconductance g_m = 34 µS/µm for V_{ds} = 4 V). Furthermore, it was demonstrated that the electrical breakdown of the monolayer MoS$_2$ was recorded at a current density of 5 × 10^7 A/cm^2, exceeding the current-carrying capacity of Cu by 50 times.

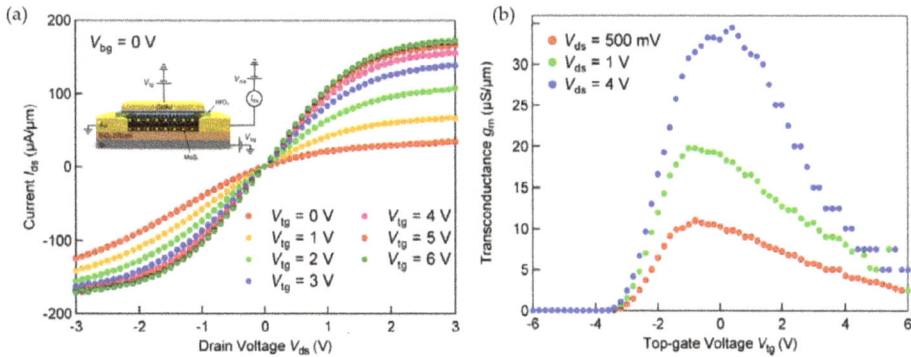

Figure 6. Performance of a high-transconductance MoS$_2$ FET. (**a**) I_{ds}–V_{ds} characteristics measured for different top-gate voltages V_{tg} from 0 to 6 V. Inset: Three-dimensional representation of the monolayer MoS$_2$ FET (**b**) Transconductance derived from I_{ds}–V_{ds} characteristics at V_{ds} = 0.5 V, 1 V, and 4 V. Reproduced with permission from [103], American Chemical Society, 2012.

The performance of monolayer TMDCs are also sensitive to random perturbations in the local environment. Current hysteresis under different ambient condition and light illumination was reported for FETs based on monolayer MoS$_2$. It was reported that current hysteresis increase steadily with the increase in humidity (RH), as shown in Figure 7a. It was proposed that such a hysteresis is due to the trapping states induced by the adsorption of water molecules on the surface of MoS$_2$. Similar hysteresis was observed when the device was under white light illumination, as shown in Figure 7b. This is attributed to the improvement of carrier concentration in the conduction channel due to photosensitivity [99]. This work indicates that controllable hysteretic behavior in MoS$_2$ FETs has the potential for humidity sensors and non-volatile memory devices.

The sensitivity of 2D TMDCs to the local environment has also contributed to flicker noise. The 1/f noise from un-encapsulated FETs based on monolayer MoS$_2$ was explained by the Hooge parameter ranging between 0.005 and 2 in vacuum (<10^{-5} Torr) [143]. The noise amplitude was reported to reduce by an order of magnitude after annealing, revealing the significant influence of atmospheric adsorbates on the charge transport [144]. On the other hand, high-k dielectric for top gate configuration is known to reduce the drain current noise in both monolayer and multilayer MoS$_2$ FETs [94,145]. Additionally, the observation of low frequency generation-recombination noise at low temperature could be due to charge traps in the underlying SiO$_2$ substrate or mid-gap states in the monolayer MoS$_2$ [143,146]. On the other hand, FETs based on layered 2H-MoTe$_2$ flakes were investigated for their low frequency noise in both vacuum and air [147]. Similar experiments were also performed using as-fabricated and aged MoS$_2$ transistors [148]. The measurements demonstrate that the flicker noise is determined by an intrinsic TMDCs conducting channel rather than the contact barriers. As a result, these noise metrics are expected to provide a unique diagnostic tool for researchers as they develop high-performance sensing applications.

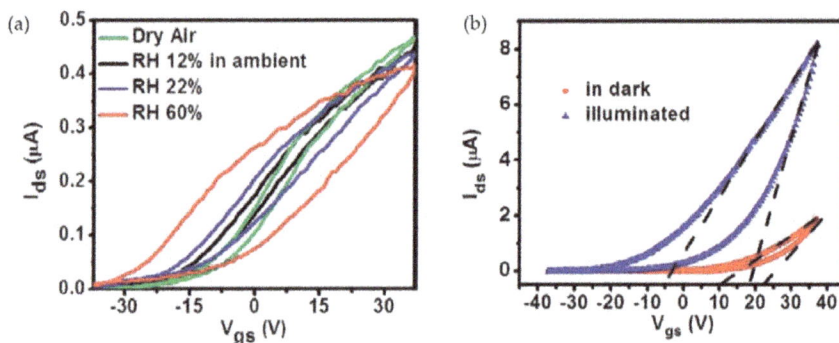

Figure 7. (**a**) Hysteresis in the monolayer MoS$_2$ transistor with controlled humidity under the sweeping rate of gate voltage at 0.5 V/s and V$_{ds}$ = 0.5 V. (**b**) Hysteresis with different illumination conditions in a monolayer MoS$_2$ transistor. Blue dots are under global white illumination (0.7 mW/cm^2), and red dots are in the dark. The intersections between the dashed lines, and y = 0 show the threshold voltages. Reproduced with permission from [99], American Chemical Society, 2012.

The effects of substrates on the performance of FETs constructed by monolayer CVD TMDCs were investigated [70,106,108]. Results suggest that these FETs offer similar electrical performance to devices based on exfoliated TMDCs. This may imply that FETs based on CVD TMDCs can be fabricated on flexible substrates, enable their applications in flexible, transparent, 2D electronic devices. The development of CVD synthesis methods for obtaining large areas of TMDCs has become important for wafer-scale device fabrication.

Finally, FETs based on monolayer TMDC alloys were also investigated. These FETs were based on an MoWSe$_2$ monolayer (Mo$_{1-x}$W$_x$Se$_2$, $0 \leq x \leq 1$), mechanically exfoliated by Scotch tape. It was demonstrated that these FETs exhibit n-type transport behavior with on/off ratios >10^5 [149].

2.3. Ambipolar Transistors with TMDCs

Ambipolar transport was demonstrated in electric double-layer transistors (EDLTs) based on thin-flake MoS$_2$ (10 nm) using an ionic liquid as the gate to reach extremely high carrier densities of 1×10^{14} cm^{-2} [150]. The charge transfer curves of such ambipolar devices are shown in Figure 8b. For EDLTs using bulk and thin MoS$_2$, the I$_{DS}$ was increased with positive V$_G$, corresponding to the behavior of an n-type semiconductor transistor. In contrast to its commonly known property as an n-type semiconductor in Table 2, when the V$_G$ turn to negative bias, an obvious p-type transport was observed to start at -1 V in the EDLT made of thin flake MoS$_2$. There was no significant change in current in the EDLT made of bulk MoS$_2$. This phenomenon indicates that hole transport is comparable to electron transport in thin-flake MoS$_2$. Hall effect measurements also revealed the mobility of 44 and 86 cm^{-1}/V^{-1}s^{-1} for electron and hole, respectively. The hole mobility is even twice the value of the electron mobility. However, the on/off ratio in this device was just >100, much lower than those reported for FETs based on monolayer MoS$_2$ [82], mainly because of the finite off-current passing through the thin flakes. The greater hysteresis at lower temperature (220 K) in Figure 8a was attributed to the slow motion of ions. A stable p–n junction was also found in the MoS$_2$ EDLTs [151].

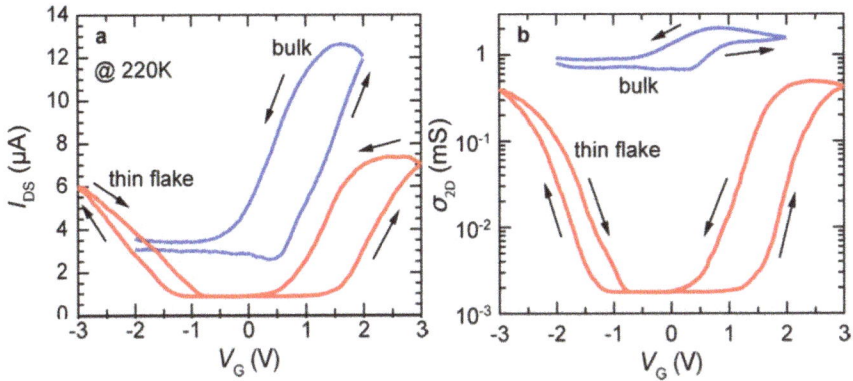

Figure 8. Transfer curve of bulk (1000 μm × 500 μm × 10 μm) and thin-flake (20 μm × 20 μm × 15 nm) MoS$_2$ EDLTs. (**a**) Change in the channel current (I_{DS}) as a function of gate voltage (V_G). (**b**) Change of sheet conductivity (σ_{2D}) as a function of gate voltage (V_G). Channel voltage (V_{DS}) is 0.2 V for both devices. Reproduced with permission from [150], American Chemical Society, 2012.

The electrical double layers (EDLs) consist of a narrow spatial charge doublet that simulates a capacitor to accumulate carriers. The use of EDLTs help to investigate the ambipolar transport in TMDCs, such as WS$_2$ [20,66], MoSe$_2$ [152], WSe$_2$ [69,116,117], and MoTe$_2$ [62]. In addition, the EDLT channel was shown to form at the interface of TMDC/ionic liquid, where the trapping state between the TMDCs and the substrates was reduced. As a result, the similar ambipolar behavior were also reported in conventional FETs based on TMDCs on dielectric materials (e.g., MoS$_2$ on PMMA [52], WS$_2$ on Al$_2$O$_3$ [65], and WSe$_2$ on hexagonal boron nitride (h-BN) [153]).

Ambipolar EDLTs based on multilayer WSe$_2$ with a high on/off ratio >10^7 at 170 K and large carrier mobility (electron mobility to 330 cm^2·V^{-1}·s^{-1} and hole mobility to 270 cm^2·V^{-1}·s^{-1}) at 77 K were reported [69]. As shown in Figure 9, it was obtained by using a low-resistance ionic liquid gate with hBN encapsulation and graphene contacts on the WSe$_2$. It was suggested that a drastic reduction of the Schottky barriers between the channel and the graphene contact electrodes by ionic liquid gating is needed in order to observe the intrinsic, phonon-limited conduction in both the electron and hole channels. However, the EDLTs still require much effort in atomic 2D-TMDC devices due to the integration, reliability, and low operating speed.

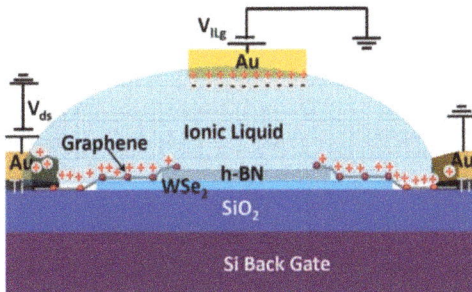

Figure 9. Schematic illustration of the structure and working principle of a hexagonal boron nitride (h-BN)/WSe$_2$ FET with ionic-liquid-gated graphene contacts. Reproduced with permission from [69], American Chemical Society, 2014.

In the case of conventional FETs, another method is using electrode metals with different work functions to control the transistor polarity. If the Fermi level of the electrode is close to the conduction band of the semiconductor, the electron injection becomes predominant. While the Fermi level is close to the valence band of the semiconductor, it translates into a hole injection. For example, due to the Fermi level of most metals close to the conduction band of MoS_2, the carrier of MoS_2 FETs were dominated by electrons [49]. With the exception of Pd [154] and Pt [131], which acted as both source and drain electrodes, MoS_2 transistors show p-type behavior. Furthermore, WSe_2 was found to exhibit ambipolar behavior by using different metals as the source and drain electrodes [155]. Alternatively, chemical doping on the surface of TMDCs (e.g., NO_2 on WSe_2 [115], K on WSe_2 [156]) by thinning the Schottky barrier width for carrier injection was also performed to study ambipolar behavior. However, the polarity of FETs was fixed after device fabrication.

The ambipolar transport were also discovered in $MoSe_2$ [56], $MoTe_2$ [58], and WSe_2 [153,157–159] under different electrostatic fields. This phenomenon was attributed to the formation of two back-to-back Schottky barriers rather than only the channel conductance. Thus, transistors can function as ambipolar transistors, and the transistor polarity can be reversed by an electric signal. A doping-free ambipolar transistor made of multilayer $MoTe_2$ is proposed in which the transistor polarity (p-type and n-type) is electrostatically controlled by dual top gates as shown in Figure 10. One of the gates was used to determine the transistor polarity, while the other gate was used to modulate the drain current [160].

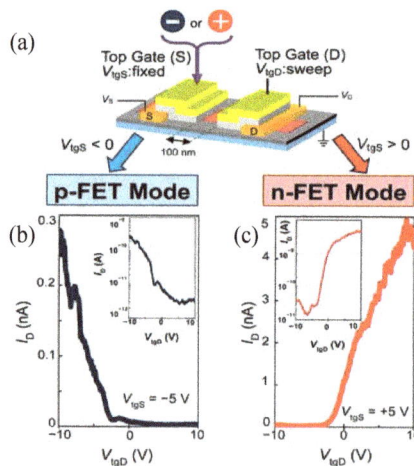

Figure 10. (**a**) Schematic of the $MoTe_2$ ambipolar transistor structure. The gap between the two top gates is 100 nm. (**b**) Experimental results for on/off operation in the p-FET mode (V_{tgS} = −5 V). The inset is a logarithmic plot. (**c**) Experimental results of an on/off operation in the n-FET mode (V_{tgS} = 5 V), with a logarithmic plot provided in the inset. Reproduced with permission from [160], American Chemical Society, 2015.

Overall, the demonstration of both n-type and p-type transport will be useful for applications that are more complicated such as CMOS logic circuits and p–n junction optoelectronic devices.

2.4. Transistors with Vertical Hetero-Structures

Due to the relatively weak van der Waals bonding between layers of TMDCs, and the absence of surface dangling bonds, the stacking of TMDCs layers and other 2D materials multilayers can energetically form heterostructures or heterojunctions [161,162]. Actual devices have been demonstrated as summarized in Table 4. For example, vertical heterojunction devices consisting of graphene (Gr) and 2D-TMDCs with appropriate bandgaps were suggested to resolve the low on/off-current ratio of graphene FETs and to achieve a high on–current density at extremely low operating voltage, which is difficult to accomplish in TMDC FETs. The use of graphene in vertically stacked devices was demonstrated as allowing electric field to penetrate through the heterostructures [163]. These vertical heterojunctions created Schottky barriers at the interfaces, where the TMDCs functioned as semiconductor instead of insulator (e.g., hBN) tunneling [164], which achieved a greater channel current while retaining a high on/off ratio [165–168].

As summarized in Table 4, vertical Gr/TMDCs heterostructures have been fabricated since 2012 [165–181]. For example, vertical Gr/MoS$_2$ FETs can achieve a high on/off ratio above 10^5 with on–current density as high as 5×10^4 A/cm^2 [171]. In addition, a similar vertical FET based on Gr/WS$_2$/Gr structure was reported with an on/off ratio as high as 10^6 even at room temperature [166]. At lower temperature (180 K), an extremely high on/off ratio of 5×10^7 was obtained in Gr/WSe$_2$ FETs [175].

Unlike n-type semiconductors, MoS$_2$, MoTe$_2$, and WSe$_2$ show p-type behavior, as summarized in Tables 2 and 3. Hence, multiple MoS$_2$/WSe$_2$ vertical heterostructures have been reported as p–n junctions [182–184]. A heterojunction based on p-type black phosphorus (BP) and n-type monolayer MoS$_2$ has also been demonstrated [185].

Although the study of transistors with vertical heterojunctions is still in its infancy, unique electrical and optical junction properties are expected to be found by combining various 2D materials with different work functions, bandgaps, and electron affinities. In fact, in addition to FETs, various TMDCs and other 2D materials heterojunction-based devices have recently been reported, including solar cells, photodetectors, LEDs, and memory devices, which will be discussed in following chapters.

Table 4. Summary of transistor performances with different vertical heterostructures. Graphene and boron nitride nanosheets are represented as Gr and hBN, respectively. APTES: (3-aminopropyl) triethoxysilane.

Materials and Structure	On/Off Ratio	On-Current Density (A/cm^2)	Off-Current Density (A/cm^2)	Source–Drain Voltage (V)	Reference
Gr/MoS$_2$/Ti	60 (290 K) 10^4 (150 K)	60	0.5	0.1	[165]
hBN/Gr/WS$_2$/Gr/Au	10^6	200	0.001	0.2	[166]
Gr/MoS$_2$/Ti/Au	10^2	150	<1	0.01	[169]
Gr/MoS$_2$/Ti/Au	10^3	5000	12.5	7	[170]
Gr/MoS$_2$/Ti/Au	10^5	5×10^4	0.1	0.5	[171]
Gr/MoS$_2$/Ti/Au	3	50	1.5	0.05	[172]
Gr/MoSe$_2$/Ti/Au	10^5	1000	0.05	0.5	[167]
Au/Gr/MoS$_2$	10^6	1000	0.0014	0.5	[173]
Gr/WSe$_2$/Provskite/Gr	10^6	<1	$<10^{-5}$	−1	[174]
Gr/MoS$_2$/Au	10^4	3000	0.3	0.5	[168]
Gr/WSe$_2$/Au	10^4	3100	0.31	0.5	[168]
APTES/Gr/WSe$_2$/Pt	3×10^4 (300 K) 5×10^7 (180 K)	11	0.001	0.1	[175]

3. Optoelectronic Devices

Optoelectronic devices are important to generate, detect, and control light, including photovoltaic devices (solar cells), light-emitting diodes (LEDs), photodetectors, and lasers. The electronic band structures of semiconductors decide for their possibility to absorb and emit light. For indirect bandgap semiconductors, additional phonons must be absorbed or emitted to conserve the momentum, which is much less efficient. Almost all MoX$_2$ and WX$_2$, TMDCs (X is a chalcogen atom) are expected to follow a similar indirect to direct bandgap transformation with decreasing layer thickness. The bandgap of these TMDCs covering an energy range of 1–2.1 eV, as shown in Table 1, are therefore especially suitable for optoelectronic applications [186]. Because of their atomically thin and strong stiffness [187], TMDCs are promising for flexible and transparent optoelectronics. As summarized in Figure 11, TMDCs offer good mobility in between those of organic materials and III–V semiconductors. A band gap near the Shockley–Quessier limit (~1.3 eV) and high mobilities are expected to lead to high-efficiency photovoltaic devices [188].

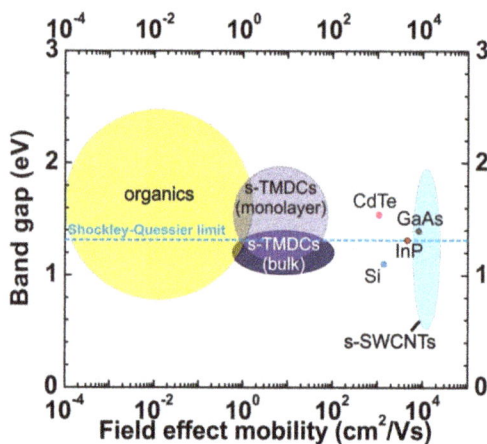

Figure 11. Bandgap vs. field-effect mobility of some important semiconductors used in optoelectronic devices. The dashed line is the band gap at ~1.34 eV while efficiency reach the Shockley-Quessier limit (~33.7%). Reproduced with permission from [188], American Chemical Society, 2014.

When irradiated with photons with an energy greater than that of the band gap, free electron/hole or bound electron–hole pairs (excitons) will be generated in the depletion region of semiconductors. The applied voltage or built-in electrical field could separate the bound excitons to generate flows of charge carriers that are called photocurrents. Hence, the p–n junction is the functional element of many optoelectronic devices, including solar cells, LEDs, and photodetectors. Devices with a combination of p-type Si and n-type MoS_2 were studied [189,190]. A p–n junction diode has been realized in MoS_2 EDLTs [151]. Similarly, the p–n junctions based on monolayer WSe_2 by electrostatic tuning were also reported [153,157–159].

3.1. Solar Cells

In Figure 12, TMDCs (MoS_2, $MoSe_2$, and WS_2) can absorb up to 5–10% incident sunlight in monolayer (thickness less than 1 nm). The sunlight absorption is one order of magnitude greater than general semiconductors such as GaAs and Si [176]. Early to 1997, Gourmelon et al. first reported the use of MoS_2 and WS_2—nanostructure absorbers in solar cells [191]. Photosensitization of TiO_2 were sensitized with WS_2 nanosheets (5 nm thick), which act as a stable absorber material [192]. Similarly, a bulk heterojunction (BHJ) solar cell made of TiO_2 nanoparticles and composites of monolayer/multilayer MoS_2 nanosheets with poly(3hexylthiophene) (P3HT) was demonstrated with 1.3% photo conversion efficiency [193]. The asymmetric Schottky junctions were demonstrated in a metal/MoS_2/metal structure, which resulted in photovoltaic devices with 0.7~2.5% power conversion efficiency (PCE) [154,194]. A greater work function difference between metals and n-type MoS_2 would generate a greater electric field in the depletion region of MoS_2. Recently, the p–n junction diodes based on TMDCs themselves were fabricated and achieved a light PCE of ~0.5%. When the device was operated as a photodiode (Figure 13), a photocurrent of 29 pA was obtained, which translates into a photo-responsivity of 16 mA/W [157]. Furthermore, the monolayer MoS_2 on p-type Si substrates formed a p–n junction and operated in tandem mode with an external quantum efficiency of 4.4% [189].

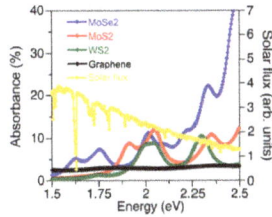

Figure 12. Bandgap absorbance of three TMD monolayers and graphene, overlapped to the incident AM1.5G solar flux. Reproduced with permission from [176], American Chemical Society, 2013.

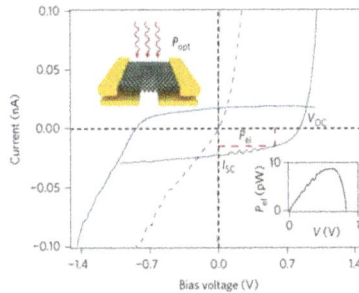

Figure 13. I–V characteristics of monolayer WSe_2 photodiode under a halogen lamp with 1400 W/m^2. The biasing conditions are p–n (solid green line; V_{G1} = −40 V, V_{G2} = 40 V), n–p (solid blue line; V_{G1} = 40 V, V_{G2} = −40 V), and p–p (dash blue line; V_{G1} = V_{G2} = 40 V). The red dashed rectangle in the main panel shows the maximum power conversion efficiency area (P_{el}). Top inset: Schematic of WSe_2 monolayer device with split gate electrodes. Lower inset: electrical power (P_{el}) versus voltage under incident illumination. Reproduced with permission from [157], Nature Publishing Group, 2014.

Highly efficient photocurrent generation was also demonstrated in vertical heterostructures. Studies on Schottky junction solar cells consisting of a graphene–MoS_2 stack suggested a power conversion efficiency of 0.1–1% [176], while p–n junctions of MoS_2/WS_2 [176] and MoS_2/WSe_2 [182] were 0.4–1.5% and 0.2%, respectively. In addition, a special vertical p–njunction made of p-type MoS_2 (CHF_3 plasma treated) and n-type MoS_2 was fabricated successfully. This MoS_2 heterostructure-based solar cell achieved up to 2.8% PCE under AM1.5G illumination [195].

However, the efficiency of solar cells based on ultrathin TMDCs is limited by the loss of absorption under the thickness limitation [196]. Calculations suggest that a monolayer of TMDCs could absorb as much sunlight as 50 nm of Si and generate electrical currents as high as 4.5 mA/cm^2 [176]. To improve the absorption of light, plasmonic enhancement technics were introduced by decorating Au nanoparticles on MoS_2 nanosheets [197]. The Au nanoparticles can produce an enhanced local optical field on an MoS_2 phototransistor device. A similar phenomenon was also investigated in the WS_2/graphene and WSe_2/graphene heterostructures, leading to enhanced photon absorption [177,178]. Alternatively, multi-junction solar cells with different bandgaps can convert different portions of the solar spectrum to reduce absorption loss [196,198]. On the other hand, a continuous bandgap tuning by strain was demonstrated [199]. A photovoltaic device made from a strain engineered MoS_2 monolayer will capture a broad range of the solar spectrum and concentrate charge carriers. These structures can also potentially be constructed using different TMDCs with varying bandgaps, which range from the visible to the near-infrared (NIR).

3.2. Light-Emitting Diodes

Light-emitting diodes (LEDs) is another type of application based on p–n junctions. Electrons and holes are separated by doping material into a p–n junction. They recombine to release energy as photons in response to electrical bias voltages. This effect is called electroluminescence (EL). Table 5 summarizes performances of LEDs constructed by various TMDCs. As direct bandgap materials (1–2 eV), monolayer 2D-TMDCs are ideal for ultrathin and flexible light-emitting layers. One of the earliest light emission studies on TMDCs was one in which MoS_2/Au nano-contacts were stimulated by scanning tunneling microscopy (STM) [200]. EL behavior was also found from exfoliated SnS_2 incorporated into a composite polymer semiconductor, but MoS_2 resulted in no light emission [201].

Table 5. Performance of light-emitting diodes (LEDs) with various transition metal dichalcogenide (TMDC) structures.

Materials and Structure	Type	EL Peak (nm)	EL Efficiency	Reference
Au/Cr/MoS$_2$/Cr/Au	Schottky junction	685	0.001%	[202]
Ti/WSe$_2$/Ti	p–i–n junction	740	0.06%	[203]
Au/Pd/WSe$_2$/Ti/Au	p–n junction	800	0.1%	[157]
Au/WSe$_2$/Pt	p–n junction	752	0.2%	[158]
Au/WSe$_2$/Au	p–n junction	750	0.1%	[159]
Ni/MoS$_2$/WSe$_2$/Au	p–n junction	800	12%	[184]
Gr/hBN/MoS$_2$/hBN/Gr		678	8.4%	
Gr/hBN/WS$_2$/hBN/Gr	Tunnel junction	620	1.32%	[204]
Gr/hBN/WSe$_2$/hBN/Gr		738	5.4%	

The result of EL in MoS_2 was reconfirmed by a monolayer MoS_2 FET, shown in Figure 14. A Schottky barrier between MoS_2 and Cr/Au was formed with a height of 100 to 400 meV. By comparing the absorption, photoluminescence (PL) and EL of the same MoS_2 monolayer, they all involved the same energy state at 1.8 eV (685 nm) [202]. WS_2 were also used to fabricate light-emitting transistors based on similar Schottky barrier between TMDCs and metals [20].

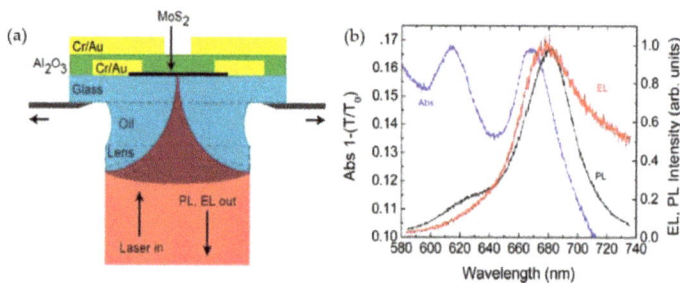

Figure 14. (a) Schematic of a top-gated MoS_2 FET and the optical setup. (b) Absorption (Abs), EL, and PL spectra on the same 1L-MoS_2. Reproduced with permission from [202], American Chemical Society, 2013.

More LEDs based on p–n junctions have been shown to have a high light emission efficiency (~1%). Homojunction-based LEDs formed by electrostatic doping of monolayer WSe_2 resulted in a maximum EL efficiency of 1% [157–159]. However, the difficulty in doping has prevented the fabrication of LEDs by monolayer TMDCs. Moreover, based on the type II heterojunction design as previous mentioned in

Section 3.1, excitonic EL was observed via a n-type monolayer MoS_2 and p-type Si heterojunction [189]. In addition, an EL spectrum in a vertically stacked MoS_2/WSe_2 heterojunction (n-type for MoS_2 and p-type for WSe_2) was also obtained with a 12% external quantum efficiency [184].

Furthermore, a new design of LEDs was introduced by stacking graphene (electrode), hexagonal boron nitride (hBN) (tunneling junction), and various semiconducting monolayers (light emitter) into complex sequences. These devices have exhibited an extrinsic quantum efficiency of nearly 10%, and the emission can be tuned over a wide range of frequencies by appropriately choosing and combining 2D TMDCs [204,205]. However, as compared to the commercial organic LEDs with an emission efficiency (external quantum efficiency) of 15–40% [206] and of ~50% in LED light bulbs, there are significant needs to further improve the performance TMDC-based LEDs. For example, more investigation on doping and surface engineering, encapsulation, and device design for band structure engineering are desired to further improve the efficiency of these LEDs [188].

Recently, LEDs have combined vertical tunnel junctions, and lateral p–n junctions have also been exploited. Such lateral LED devices were constructed by two graphene transparent electrodes, two hBN tunnel barriers, and one monolayer WSe_2. Two separated back-gates were utilized to electrostatically define the p–i–n junction in WSe_2 [203,207]. A single photon emitter in a WSe_2 LED based on a p–i–n junction was reported; a narrow EL peak (width < 300 μeV) at 1.704 eV (728 nm) under low current (~100 nA) and low temperature (~5 K) was observed [207].

3.3. Photodetectors

Photodetectors are light sensors with p–n junctions that convert photons into electrical current. The junction is usually covered by a window with anti-reflective coatings. The absorption of photons will create electron–hole pairs in the depletion region. Photodiodes and phototransistors are two major examples of such photodetectors. In terms of frequency response, photodiodes are much faster than phototransistors (ns vs. μs) and not sensitive to temperature fluctuation. Therefore, photodiodes usually work as solar cells and photoconductors. Phototransistors are transistors with the base terminal exposed instead of sending photocurrent into the base. Compared to photodiodes, phototransistors have a greater gain.

The application of TMDCs in photodetectors has been widely demonstrated. Basically, the metal–semiconductor–metal (M-S-M) structured photodetector is made of MoS_2 and amorphous silicon with response times of about 0.3 ms and a photoresponsivity of 210 mA/W when irradiated with green light [190]. By using a two-pulse photo-voltage correlation technique, an ultrafast response of 3 ps was recorded in photodetectors constructed with monolayer MoS_2 [208]. A phototransistor made from mechanical exfoliated monolayer MoS_2 has also exhibited high photo-responsivity of 7.5 mA/W. By improved mobility and contact quality, an ultrasensitive monolayer MoS_2 photodetector was also fabricated [209]. The maximum external photo-responsivity was able to attain up to 880 A/W. As shown in Figure 15a, the drain current I_{ds} can increase upon light illumination. This device also showed a uniform increase in photo-responsivity as the illumination wavelength was reduced from 680 to 400 nm (Figure 15b). This suggests that monolayer MoS_2 photodetectors can be used for a broad range of wavelengths, between 400 and 650 nm. Phototransistors based on CVD-grown monolayer MoS_2 and multilayer WS_2 have also been demonstrated [210,211]. Photodetectors with high photo-responsivity (2.5×10^7 A/W) and external quantum efficiency (3168%) were fabricated by multilayer ReS_2 [75,78].

As it is a classic structure of TMDC-based photodetectors, the M-S-M configuration has been systematically studied. Schottky barrier modulation between 2D-TMDCs and electrodes is one of the promising methods of improving performance [212–214]. Phase transformation of MoS_2 (2H-1T) will also reduce the native Schottky barrier and increase photo-responsivity by more than one order of magnitude [215]. Self-assembled doping is another method of improving mobility and photo-responsivity [216]. After all, the photo-responsivity of the TMDC photodetectors

(~880 mA/W) [209] is much greater than that of graphene photodetectors (15.7 mA/W) [217], and the general response speed (~3 ps on MoS₂) [208] is relatively high (~50 ps on graphene) [218].

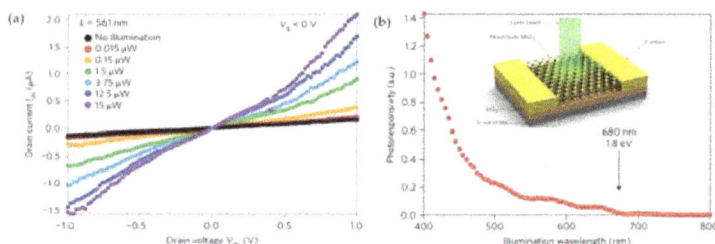

Figure 15. Performance of a monolayer MoS₂ phototransistor. (a) I_{ds}–V_{ds} characteristic of the device in the dark and under different illumination intensities. (b) Photoresponsivity of a similar monolayer MoS₂ device as a function of illumination wavelength. Inset: Three-dimensional schematic view of the monolayer MoS₂ photodetector and the focused laser beam used to probe the device. Reproduced with permission from [209], Nature Publishing Group, 2013.

The photothermoelectric effect also plays a considerable role in photo-responsivity [219]. In addition to photo-excited electron–hole pairs across the Schottky barriers at the TMDC/electrode interface, the carrier is also excited by light illumination based on temperature difference. The photovoltaic effect interface dominates in the accumulation regime, whereas the hot-carrier-assisted photothermoelectric effect prevails in the depletion regime for the multilayer MoS₂ photodetector [220]. Meanwhile, the photocurrent generation in monolayer MoS₂ is dominated by the photothermoelectric effect [221]. The photothermoelectric effect has also been studied in other TMDC-based photodetectors [153,222].

By using MoS₂ layers of different thicknesses, phototransistors can be tuned to absorb lights at different wavelengths, as shown in Figure 16 [198]. Furthermore, multilayer MoS₂ phototransistors have been demonstrated to response spectra range from ultraviolet (UV) to near-infrared (NIR) [223]. Similar bandgap tuning, by varying the number of TMDC layers, has also been applied for WS₂ [224].

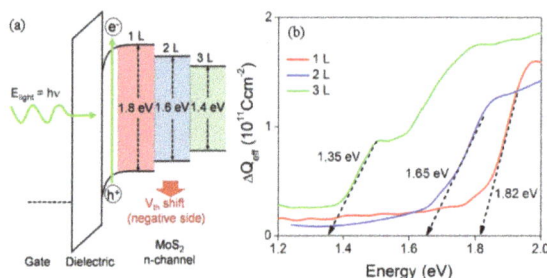

Figure 16. (a) The schematic band diagrams of ITO (gate)/Al₂O₃ (dielectric)/single-, double-, and triple-layer MoS₂ (n-channel) under the gate bias for carrier (electron–hole) accumulation. (b) Photon energy-dependent (ΔQ_eff) plots indicate the approximate optical energy gaps to be 1.35 eV, 1.65 eV, and 1.82 eV for trilayer, bilayer, and monolayer MoS₂ nanosheets, respectively. Reproduced with permission from [198], American Chemical Society, 2012.

Heterojunction fabrication can be performed to optimize the performance of 2D-TMDC photodetectors. Highly efficient photocurrent generation has been demonstrated in vertical heterostructures of Gr/TMDCs and Gr/hBN/TMDCs [177,225], where a maximum external quantum efficiency was 55%, internal quantum efficiency was up to 85%, and the response speed was down to 53.6 µs [179]. The hBN/Gr/WSe$_2$ vertical heterostructure photodetectors have a superfast photo-response time of 5.5 ps [180]. Then, a photodetector with a WSe$_2$/Gr/MoS$_2$ heterostructure, which extends the absorption spectral range from 400 to 2400 nm [181].

In addition, atomically 2D-TMDCs still have physical limitations with light absorption, so the use of an absorption layer (e.g., R6G/PbS/perovskite) was applied on 2D-TMDC photodetectors [226–228].

3.4. Lasers

In photoluminescence (PL) devices, light emission will occur after absorbing photons at greater energy. The enhancement of PL quantum yield has been demonstrated in TMDC monolayers [16,229–232], although the yield difference between bulk and monolayer MoS$_2$ was relatively modest. The absolute PL intensity of the WS$_2$ and WSe$_2$ monolayers was 20–40 times greater than that of monolayer MoS$_2$ exfoliated from a natural crystal [22,231]. Compared to MoS$_2$, WS$_2$ and WSe$_2$ are better candidates for light emission devices.

However, the spontaneous emission efficiency of monolayer TMDCs is still low because the non-radiative recombination rate exceeds the spontaneous emission rate [233]. For monolayer MoS$_2$ on SiO$_2$ substrates, the non-radiative decay time is ~70 ps and the spontaneous emission lifetime is ~10 ns, as estimated at room temperature [234,235].

The enhancement of spontaneous emission could be achieved by two major strategies as shown in Figure 17a,b. One is to couple the TMDCs with a low Q-factor (~300) planar photonic crystal (PPC) [236,237]. The other one is to integrate TMDCs with distributed Bragg reflector (DBR) micro-cavities [238,239]. Recently, device performance based on monolayer WSe$_2$ was improved by using the former strategy. The as-fabricated PCCs had Q-factors of about 10^4. This led to a significant improvement of the Purcell factor, which is crucial for lasing. The optical pumping threshold was as low as 27 nW at 130 K [240].

Besides the Purcell enhancement of spontaneous emission, there are several critical challenges for achieving lasing using 2D TMDCs. For example, the large effective masses of charge carriers in MoS$_2$ result in high densities of states in both valence and conduction bands. MoS$_2$ with 1.8 eV bandgap requires carrier concentrations to push the quasi-Fermi levels into the corresponding bands to achieve population inversion [235]. The transition between the valence band and the conduction band-edge has a strong excitonic feature in such a monolayer TMDC system, including neutral and red-shifted charged excitons, allowing for the long-lived population inversion required, which has been studied in MoS$_2$ [241], WS$_2$ [242], and WSe$_2$ [243].

In addition, the limited gains and the lack of optical confinement and feedback within the monolayer TMDCs will hinder coherent light emission. There have been demonstrations that a strong optical confinement and an enhanced modal gain can be achieved by embedding 2D TMDCs at the interface between a free-standing microsphere and a microdisk [235,244], as shown in Figure 17c,d. The devices exhibited multiple resonant lasing peaks in the wavelength range of ~575–775 nm. In Figure 17d, the structure of an Si$_3$N$_4$/WS$_2$/HSQ microdisk with a diameter of 3.3 µm has two advantages: enhanced optical mode overlap and material protection. The lasing performance did not decay even after one year, as the sandwich structure protects the monolayer from direct exposure to air [244].

It was noted that the key to the lasing lies in the monolayer of the gain medium, which confines direct-gap excitons to within micro- or even nano-structures. The configuration allows for gain properties via external controls such as electrostatic gating and current injection, enabling optically or electrically pumped operation. These results show that the possibilities of these fabrication methods are scalable or designable, and compatible with integrated electronic circuits technologies. However, the performance of the device might be further improved by exploring surface passivation, doping, and strain engineering.

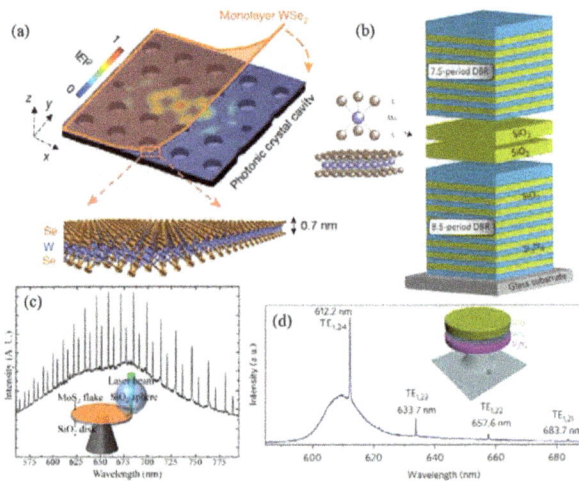

Figure 17. (**a**) Illustration of PPC architecture, where the electric-field profile (in-plane, x-y) of the fundamental cavity mode (pristine cavity before WSe$_2$ transfer) is embedded as the color plot. Reproduced with permission from [240], Nature Publishing Group, 2015. (**b**) Schematic of the DBR microcavity structure, a monolayer of MoS$_2$, sandwiched between two SiO$_2$ layers to form the cavity layer. Reproduced with permission from [239], Nature Publishing Group, 2014. (**c**) PL spectrum of the laser device with an excitation power of 30 mW at room temperature. Inset: Schematic configuration of the coupled SiO$_2$ microsphere/microdisk optical cavity with the incorporation of MoS$_2$ flake. Reproduced with permission from [235], American Chemical Society, 2015. (**d**) PL spectrum taken at 10 K when the pump intensity is above lasing threshold, showing whispering gallery modes (WGMs) at 612.2 nm, 633.7 nm, 657.6 nm, and 683.7 nm. Inset: Schematic image of the Si$_3$N$_4$/WS$_2$/HSQ micro sphere sandwich structure. Reproduced with permission from [244], Nature Publishing Group, 2015.

4. Integrated Circuits

4.1. Amplifiers and Inverters

An amplifier is a basic FET device that utilizes power from the supply to amplify the input signals to greater amplitudes. The operation of an analog small signal amplifier was demonstrated based on monolayer MoS$_2$ [102]. As shown in Figure 18a, the device consisted of two transistors that were integrated on a single monolayer MoS$_2$. The gate of one transistor (the switch) acted to provide input, while the gate of the other transistor (the load) was connected to the central lead and acted as the output/amplifier. The power supply of the amplifier V$_{DD}$ was set to 2 V. The switch and load transistors connected in series are represented in Figure 18b. The switch transistor was first biased at a certain DC voltage to establish a desired drain current, shown as the Q-point (quiescent point) in Figure 18c. When a small sinusoidal AC signal V$_{in-AC}$ of amplitude $\Delta V_{in}/2$ was superimposed on the DC signal at the input, the output voltage ΔV_{out} oscillated synchronously with a phase difference of 180° with respect to V$_{in-AC}$. The slope of the red straight line in Figure 18c represented the gain

$$G = \frac{\Delta V_{out}}{\Delta V_{in}} > 4$$

of the amplifier at room temperature. By increasing the frequency of the AC signal, the gain was reduced to 1 at 2000 Hz, as shown in Figure 18d.

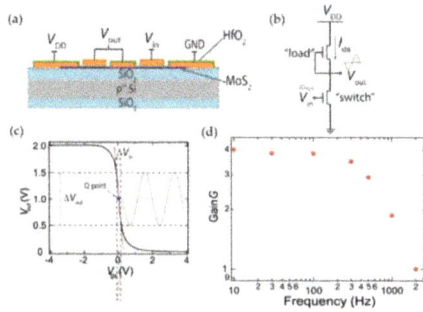

Figure 18. (**a**) Vertical cross section of the MoS$_2$ amplifier device. (**b**) Schematic drawing of integrated amplifier in common-source configuration. (**c**) Transfer characteristic of the integrated amplifier realized with two transistors on the same MoS$_2$ flake. (**d**) Voltage gain dependence on frequency of the small input signal. Reproduced with permission from [102], American Institute of Physics, 2012.

An important application of amplifiers is logic gate inverters (NOT gates). A NOT gate is designed to convert Logical 0 (low input voltage) to Logical 1 (high output voltage), and vice versa. The first monolayer MoS$_2$ inverter was reported in 2011 [97]. Up to six independently switchable transistors were fabricated on a single monolayer MoS$_2$ flake by lithographically patterning. As shown in Figure 19, an NOR-gate logic operation was also demonstrated, making them suitable for incorporation into digital circuits. A NOR gate is a universal gate that can be built in combinations to form all other logic operations.

On the other hand, radio frequency (RF) amplifiers and a logic inverter can be formed by integrating multiple MoS$_2$ transistors on quartz or flexible substrates. These devices have been demonstrated with an intrinsic gain over 30 and work in the GHz regime [90]. Several reports on amplifiers or inverters based on TMDCs were also reported with a voltage gain ranging from 1.45 to 27 [58,91,97,116,165,245–248]. Low-power consumption complementary inverters were also fabricated in the sub-nanowatt range [248], which is the most important advantage of CMOS inverters over single FET-type inverters.

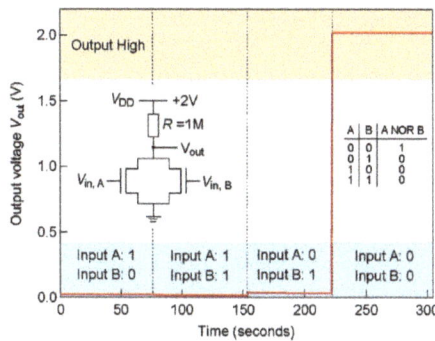

Figure 19. Demonstration of an NOR-gate logic circuit based on monolayer MoS$_2$ transistors. The output voltage V$_{out}$ is shown for four different combinations of input states (1,0), (1,1), (0,1), and (0,0). The output is in the high state only if both inputs are in the low state. Left inset: The circuit is formed by connecting two monolayer MoS$_2$ transistors in parallel and using an external 1 MΩ resistor as a load. Reproduced with permission from [97], American Chemical Society, 2011.

4.2. Logic Circuits

Besides amplifier and inverters, more complicated devices of logic circuits have also been investigated. Wang et al. demonstrated complex integrated circuits based on bilayer MoS_2. These ICs consist of inverters, logical NAND gates, a static random access memory (SRAM), and a five-stage ring oscillator [249]. It is noted that an NAND gate is another basic logic gate with universal functionality. Any other type of logic gate can then be constructed with a combination of NAND gates. The schematic design and the optical micrograph of an NAND gate and SRAM circuit fabricated on the bilayer MoS_2 is shown in Figure 20. The SRAM was constructed from a pair of cross-coupled inverters as shown. This storage cell had two stable states (0 and 1) at the output. The flip-flop cell could be set to Logic State 1 (or 0) by applying a low (or high) voltage to the input.

The five-stage ring oscillator (Figure 21A,B) was constructed to assess the high frequency switching capability and for evaluating MoS_2 ultimate compatibility with conventional circuit architecture. The positive feedback loop in the ring oscillator resulted in a statically unstable system, and the voltage at the output of each inverter stage oscillated as a function of time as shown in Figure 21C. The output signal of the ring oscillator can also be measured in terms of its frequency power spectrum. As shown in Figure 21D, the fundamental oscillation frequency was at 1.6 MHz at $V_{dd} = 2$ V, corresponding to a propagation delay of 62.5 ns per stage [249].

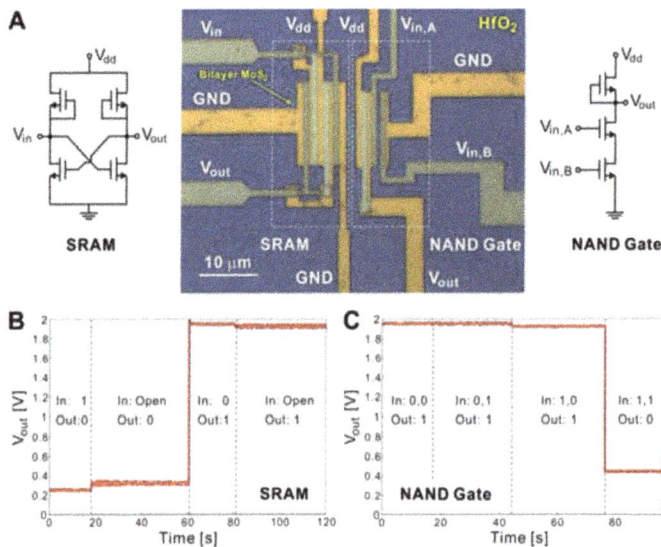

Figure 20. Demonstration of an integrated NAND logic gate and a static random-access memory (SRAM) cell on bilayer MoS_2. (**A**) Optical micrograph of the NAND gate and the SRAM fabricated on the same bilayer MoS_2 thin film. The corresponding schematics of the electronic circuits for the NAND gate and SRAM are also shown. (**B**) Output voltage of the flip-flop memory cell (SRAM). A Logic State 1 (or 0) at the input voltage can set the output voltage to Logic State 0 (or 1). In addition, the output logic state stays at 0 or 1 after the switch to the input has been opened. (**C**) Output voltage of the NAND gate for four different input states: (0,0), (0,1), (1,0), and (1,1). A low voltage below 0.5 V represents Logic State 0 and a voltage close to 2 V represents Logic State 1. Reproduced with permission from [249], American Chemical Society, 2012.

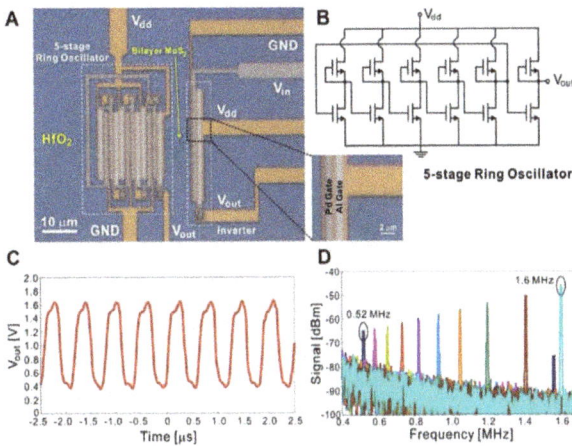

Figure 21. A five-stage ring oscillator based on bilayer MoS$_2$. (**A**) Optical micrograph of the ring oscillator constructed on a bilayer MoS$_2$ thin film. (**B**) Schematic of the electronic circuit of the five-stage ring oscillator. The first five inverter stages form the positive feedback loop, which leads to the oscillation in the circuit. The last inverter serves as the synthesis stage. (**C**) Output voltage as a function of time for the ring oscillator at V_{dd} = 2 V. (**D**) The power spectrum of the output signal as a function of V_{dd}. From left to right, V_{dd} = 1.15 V and 1.2 to 2.0 V in steps of 0.1 V. The corresponding fundamental oscillation frequency increases from 0.52 to 1.6 MHz. Reproduced with permission from [249], American Chemical Society, 2012.

4.3. Memory Devices

Memory cells are more basic building blocks of integrated digital electronics. In fact, graphene-based nonvolatile memory devices were studied earlier because of the high carrier mobility. However, graphene memory devices typically exhibited very low program/erase (P/E) current ratios up to 5.5 due to the absence of an energy bandgap. To overcome this limitation in graphene-based memory devices, TMDCs with a sufficient bandgap was proposed as a channel material for nonvolatile memory devices. The TMDC FETs showed a high on/off-current ratio and have the potential to enhance the P/E ratio in nonvolatile memory systems. The first nonvolatile memory cell based on TMDCs was demonstrated in 2012 with monolayer MoS$_2$ [101]. The P/E ratio can reach 5×10^3 at 1 V, as shown in Table 6.

Table 6. Performance of memory devices based on TMDCs and other materials.

Materials	P/E Ratio	Endurance (Cycles)	Retention (s)	Operating Voltage(V)	Reference
MoS$_2$ (with PZT)	5×10^3		10^3	1	[101]
	10^4	500	10^4	0.1	[250]
	10^4	120	2×10^3	0.05	[251]
Graphene/MoS$_2$	10^4	120	2×10^3	0.05	[252]
Graphene/hBN/MoS$_2$	10^5	120	1.4×10^3	0.05	[253]
	10^5	10^5	10^4	8	[254]
MoS$_2$/hBN/BP	50	40	10^3	0.05	[113]

Furthermore, nonvolatile memory devices that combine graphene with monolayer MoS$_2$, which reduced the operation voltage (V_{ds}) to 0.05 V, have been demonstrated [252]. As shown in Figure 22a, monolayer MoS$_2$ is used as the semiconducting channel with graphene stripes as the electrodes in FET geometry. A piece of multilayer graphene (MLG), separated by a 6-nm-thick tunneling

HfO$_2$ layer from the monolayer MoS$_2$ channel was used as the floating gate. The conductivity of the MoS$_2$ channel was modulated by the voltage V$_{CG}$ applied on the top Cr/Au control gate electrode (CG). When a positive V$_{CG}$ is applied to the device, electrons will tunnel from the MoS$_2$ channel through the HfO$_2$ and accumulate on the MLG floating gate. The application of a negative V$_{CG}$ will deplete the floating gate and reset the device, as shown in Figure 22b. The MLG has a work function of 4.6 eV, which is not sensitive to the number of layers and results in a deep potential well for charge trapping, thus improving the charge retention. Monolayer MoS$_2$ was highly sensitive to the presence of charges in the floating gate MLG, resulting in a 10^4 difference between memory program and erase states in Figure 22c.

In addition, a new type of the memory device was fabricated with graphene (G) as the FET channel, hBN (B) as the tunnel barrier, and MoS$_2$ (M) as the charge trapping layer (denoted as GBM) [253,254]. This result confirmed that the MoS$_2$ layer could act as an effective charge-trapping layer. These GBM devices also showed different hysteresis characteristics, depending on the thicknesses of MoS$_2$ and hBN. When a thicker MoS$_2$ layer and thinner hBN were employed, unipolar conductance and greater hysteresis were observed due to effective electron tunneling and electric-field screening. In addition, the reversed stacking structure (MBG) was also investigated, with a high on/off current ratio and a large memory window [253]. Due to the time-dependent PL, the MoS$_2$–graphene heterostructure can also function as a rewritable optoelectronic switch or memory, where the persistent state shows almost no relaxation or decay within experimental timescales, indicating near-perfect charge retention [255].

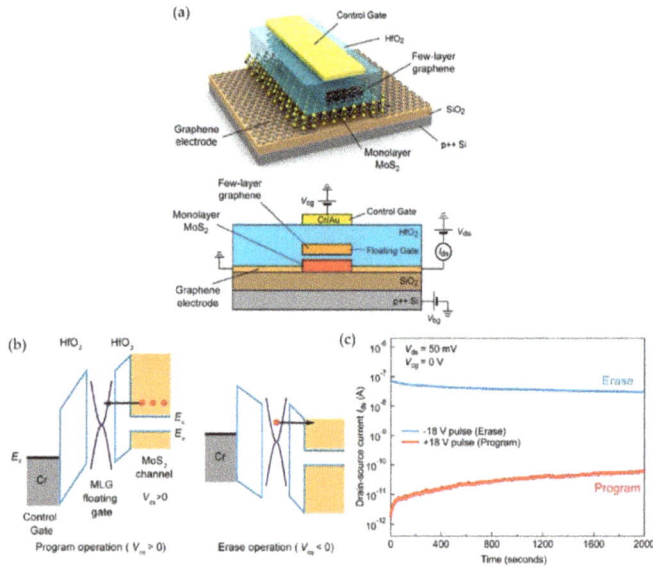

Figure 22. MoS$_2$/graphene heterostructure memory device layout. (**a**) Three-dimensional schematic view of the memory device based on single-layer MoS$_2$. Lower is the schematics of a heterostructure memory cell with a single-layer MoS$_2$ semiconducting channel, graphene contacts and MLG floating gate. The MLG floating gate is separated from the channel by a thin tunneling oxide (1 nm Al$_2$O$_3$ + 6 nm HfO$_2$) and from the control gate by a thicker blocking oxide (1 nm Al$_2$O$_3$ + 30 nm HfO$_2$). (**b**) Simplified band di+agram of the Cr/HfO$_2$/Gr/HfO$_2$/MoS$_2$ heterostructure in program and erase operation. (**c**) Temporal evolution of drain–source currents (I$_{ds}$) in the erase (ON) and program (OFF) states. The curves are acquired independently for the program (10^{-8}–10^{-7} A) and erase (10^{-12}–10^{-10} A) current states and plotted on a common time scale. The drain–source bias voltage (V$_{ds}$) is 50 mV, and the duration of the control-gate voltage (V$_{cg}$) pulse is 3s. Reproduced with permission from [252], American Chemical Society, 2013.

The combination of TMDCs with ferroelectric materials was also used for nonvolatile memory devices as schematically drawn in Figure 23 [101]. FETs based on monolayer or multilayer MoS_2 were fabricated on a lead zirconium titanate ($Pb(Zr,Ti)O_3$, PZT) substrates. The PZT substrate can be used as the alternative to the optoelectronic switching, to write and erase the data as shown in Figure 23. When the MoS_2 layer was illuminated by visible light, the photo-generated carrier would create an internal electric field to affect the polarization of PZT beneath the MoS_2. Only after 5 min illumination, the ON and OFF states became indistinguishable, and data was erased [250].

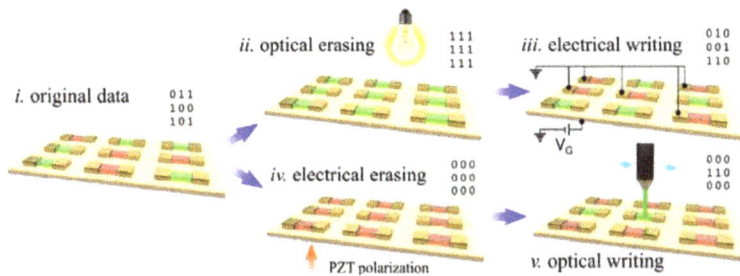

Figure 23. "Optical erase-electrical write" and "electrical erase-optical write" operation of MoS_2-PZT memories. Reproduced with permission from [250], American Chemical Society, 2015.

A type-switching memory device formed with a BP/hBN/MoS_2 heterostructure was developed based on the ambipolar property of BP memory devices [256], and the memory device operated under p-type and n-type modes according to the polarity of a gate voltage pulse, but the P/E ratio was only up to 50 [113], which is not comparable to the Gr/MoS_2-based devices. These MoS_2-based memory devices exhibited high-performance in terms of P/E ratio (10^5), long endurance, and retention (10^5 cycles) and have a low operating voltage (0.05 V). It is expected that TMDCs can be used for non-volatile, portable, and power saving memory devices.

5. Summary

In this review, we have critically evaluated the progress of electronic and optoelectronic devices based on 2D TMDCs. Apparently, the application of TMDCs for field effect transistors (FETs) has been very popular. Most efforts have focused on improving the device mobility by various approaches, including encapsulation with dielectric films, gate configurations, managing the contact barriers, and using "defect-free" substrates such as BN nanosheets. The progress in electronic devices has now involved the development of complex integrated circuits (ICs) and logic gates. Many of these are based on the layout design of the circuit on a same piece of 2D TMDC, fully demonstrating the advantage of "large-area" 2D materials for device integration. On the other hand, many advances have been demonstrated for optoelectronic devices, including solar cells, light-emitting diodes (LEDs), photodetectors, and even lasers. Overall, the efficiency of the TMDC-based optoelectronics is still low. Some of the major issues have been the difficulty in p-type and n-type doping on 2D TMDCs. Other issues are due to the atomically thin nature of the materials, which prevent sufficient light absorption, in spite of their high absorption coefficient (~10^7 m^{-1} in the visible range) [257]. "Stacked heterostructures" appear to be a potential solution, either by stacking multiple 2D TMDCs or other 2D materials (graphene and BN nanosheets), or stacking on other functional "substrates" such as photonic structures, among others. However, most of the 2D heterostructure devices were fabricated by mechanical exfoliation and transference, which is complicated and inefficient for opto-electro applications. The improvement of 2D heterostructure synthesis on a large scale is critical for developing the 2D materials.

It is noted that TMDCs are among the very few semiconducting 2D materials that are stable for device applications. The semiconducting properties in TMDCs are the important features that supplement the gapless graphene and the electrically insulating hBN for 2D devices. The emergence of TMDCs has enabled meaningful applications of vertically stacked heterostructures for novel electronics and optoelectronics devices with tailorable properties. Overall, the progress on 2D TMDC devices is still in its infancy and requires further development. The biggest issue of the TMDCs is the relatively low performance of their electronic and optoelectronic devices, as compared to those constructed by other state-of-the-art semiconductors. The major advantages of TMDC-based devices are the use of very few materials (atomically thin) and the capability of being fabricated as flexible and wearable devices. TMDCs enable energy/materials saving fabrication of highly portable devices, which meet the desired features of nanotechnology.

Acknowledgments: Yoke Khin Yap acknowledges the support from the National Science Foundation (Award number DMR-1261910).

Author Contributions: M.Y. wrote the manuscript with the supervision from D.Z. and Y.K.Y. D.Z. and Y.K.Y. plan, draft, edit and rewrite the manuscript.

Conflicts of Interest: The authors declare no conflict of interest. The founding sponsors had no role in the design of the study; in the collection, analyses, or interpretation of data; in the writing of the manuscript; or in the decision to publish the results.

References

1. Lundstrom, M. Applied physics. Moore's law forever? *Science* **2003**, *299*, 210–211. [CrossRef] [PubMed]
2. Theis, T.N.; Solomon, P.M. It's time to reinvent the transistor! *Science* **2010**, *327*, 1600–1601. [CrossRef] [PubMed]
3. Morton, J.J.; McCamey, D.R.; Eriksson, M.A.; Lyon, S.A. Embracing the quantum limit in silicon computing. *Nature* **2011**, *479*, 345–353. [CrossRef] [PubMed]
4. Ferain, I.; Colinge, C.A.; Colinge, J.P. Multigate transistors as the future of classical metal-oxide-semiconductor field-effect transistors. *Nature* **2011**, *479*, 310–316. [CrossRef] [PubMed]
5. Ionescu, A.M.; Riel, H. Tunnel field-effect transistors as energy-efficient electronic switches. *Nature* **2011**, *479*, 329–337. [CrossRef] [PubMed]
6. Lee, C.H.; Qin, S.; Savaikar, M.A.; Wang, J.; Hao, B.; Zhang, D.; Banyai, D.; Jaszczak, J.A.; Clark, K.W.; Idrobo, J.C.; et al. Room-Temperature Tunneling Behavior of Boron Nitride Nanotubes Functionalized with Gold Quantum Dots. *Adv. Mater.* **2013**, *25*, 4544–4548. [CrossRef] [PubMed]
7. Hao, B.; Asthana, A.; Hazaveh, P.K.; Bergstrom, P.L.; Banyai, D.; Savaikar, M.A.; Jaszczak, J.A.; Yap, Y.K. New Flexible Channels for Room Temperature Tunneling Field Effect Transistors. *Sci. Rep.* **2016**, *6*. [CrossRef] [PubMed]
8. Parashar, V.; Durand, C.P.; Hao, B.; Amorim, R.G.; Pandey, R.; Tiwari, B.; Zhang, D.; Liu, Y.; Li, A.P.; Yap, Y.K. Switching Behaviors of Graphene-Boron Nitride Nanotube Heterojunctions. *Sci. Rep.* **2015**, *5*, 12238. [CrossRef] [PubMed]
9. Shim, J.; Park, H.-Y.; Kang, D.-H.; Kim, J.-O.; Jo, S.-H.; Park, Y.; Park, J.-H. Electronic and Optoelectronic Devices based on Two-Dimensional Materials: From Fabrication to Application. *Adv. Electron. Mater.* **2017**, *3*, 1600364. [CrossRef]
10. Ye, M.; Winslow, D.; Zhang, D.; Pandey, R.; Yap, Y. Recent Advancement on the Optical Properties of Two-Dimensional Molybdenum Disulfide (MoS_2) Thin Films. *Photonics* **2015**, *2*, 288–307. [CrossRef]
11. Tongay, S.; Zhou, J.; Ataca, C.; Lo, K.; Matthews, T.S.; Li, J.; Grossman, J.C.; Wu, J. Thermally driven crossover from indirect toward direct bandgap in 2D semiconductors: $MoSe_2$ versus MoS_2. *Nano Lett.* **2012**, *12*, 5576–5580. [CrossRef] [PubMed]
12. Zhang, Y.; Chang, T.R.; Zhou, B.; Cui, Y.T.; Yan, H.; Liu, Z.; Schmitt, F.; Lee, J.; Moore, R.; Chen, Y.; et al. Direct observation of the transition from indirect to direct bandgap in atomically thin epitaxial $MoSe_2$. *Nat. Nanotechnol.* **2014**, *9*, 111–115. [CrossRef] [PubMed]
13. Goldberg, A.M.; Beal, A.R.; Lévy, F.A.; Davis, E.A. The low-energy absorption edge in $2H-MoS_2$ and $2H-MoSe_2$. *Philos. Mag.* **1975**, *32*, 367–378. [CrossRef]

14. Coehoorn, R.; Haas, C.; Dijkstra, J.; Flipse, C.J.; de Groot, R.A.; Wold, A. Electronic structure of MoSe$_2$, MoS$_2$, and WSe$_2$. I. Band-structure calculations and photoelectron spectroscopy. *Phys. Rev. B Condens. Matter* **1987**, *35*, 6195–6202. [CrossRef] [PubMed]

15. Ma, Y.; Dai, Y.; Guo, M.; Niu, C.; Lu, J.; Huang, B. Electronic and magnetic properties of perfect, vacancy-doped, and nonmetal adsorbed MoSe$_2$, MoTe$_2$ and WS$_2$ monolayers. *Phys. Chem. Chem. Phys.* **2011**, *13*, 15546–15553. [CrossRef] [PubMed]

16. Ruppert, C.; Aslan, O.B.; Heinz, T.F. Optical properties and band gap of single- and few-layer MoTe$_2$ crystals. *Nano Lett.* **2014**, *14*, 6231–6236. [CrossRef] [PubMed]

17. Grant, A.J.; Griffiths, T.M.; Pitt, G.D.; Yoffe, A.D. The electrical properties and the magnitude of the indirect gap in the semiconducting transition metal dichalcogenide layer crystals. *J. Phys. C Solid State Phys.* **1975**, *8*, L17–L23. [CrossRef]

18. Hind, S.P.; Lee, P.M. KKR calculations of the energy bands in NbSe$_2$, MoS$_2$ and alpha MoTe$_2$. *J. Phys. C Solid State Phys.* **1980**, *13*, 349–357. [CrossRef]

19. Ding, Y.; Wang, Y.; Ni, J.; Shi, L.; Shi, S.; Tang, W. First principles study of structural, vibrational and electronic properties of graphene-like MX2 (M = Mo, Nb, W, Ta; X = S, Se, Te) monolayers. *Phys. B Condens. Matter* **2011**, *406*, 2254–2260. [CrossRef]

20. Jo, S.; Ubrig, N.; Berger, H.; Kuzmenko, A.B.; Morpurgo, A.F. Mono- and bilayer WS$_2$ light-emitting transistors. *Nano Lett.* **2014**, *14*, 2019–2025. [CrossRef] [PubMed]

21. Kam, K.K.; Parklnclon, B.A. Detailed Photocurrent Spectroscopy of the Semiconducting Grouping VI Transition Metal Dichalcogenides. *J. Phys. Chem.* **1982**, *86*, 463–467. [CrossRef]

22. Zhao, W.; Ghorannevis, Z.; Chu, L.; Toh, M.; Kloc, C.; Tan, P.-H.; Eda, G. Evolution of Electronic Structure in Atomically Thin sheets of WS$_2$ and WSe$_2$. *ACS Nano* **2013**, *7*, 791–797. [CrossRef] [PubMed]

23. Dawson, W.G.; Bullett, D.W. Electronic-Structure and Crystallography of MoTe$_2$ and Wte$_2$. *J. Phys. C Solid State Phys.* **1987**, *20*, 6159–6174. [CrossRef]

24. Tongay, S.; Sahin, H.; Ko, C.; Luce, A.; Fan, W.; Liu, K.; Zhou, J.; Huang, Y.S.; Ho, C.H.; Yan, J.; et al. Monolayer behaviour in bulk ReS$_2$ due to electronic and vibrational decoupling. *Nat. Commun.* **2014**, *5*, 3252. [CrossRef] [PubMed]

25. Friemelt, K.; Kulikova, L.; Kulyuk, L.; Siminel, A.; Arushanov, E.; Kloc, C.; Bucher, E. Optical and photoelectrical properties of ReS$_2$ single crystals. *J. Appl. Phys.* **1996**, *79*, 9268–9272. [CrossRef]

26. Wolverson, D.; Crampin, S.; Kazemi, A.S.; Ilie, A.; Bending, S.J. Raman Spectra of Monolayer, Few-Layer, and Bulk ReSe$_2$: An Anisotropic Layered Semiconductor. *ACS Nano* **2014**, *8*, 11154–11164. [CrossRef] [PubMed]

27. Yang, S.; Tongay, S.; Li, Y.; Yue, Q.; Xia, J.B.; Li, S.S.; Li, J.; Wei, S.H. Layer-dependent electrical and optoelectronic responses of ReSe$_2$ nanosheet transistors. *Nanoscale* **2014**, *6*, 7226–7231. [CrossRef] [PubMed]

28. Yang, S.; Wang, C.; Sahin, H.; Chen, H.; Li, Y.; Li, S.S.; Suslu, A.; Peeters, F.M.; Liu, Q.; Li, J.; et al. Tuning the optical, magnetic, and electrical properties of ReSe$_2$ by nanoscale strain engineering. *Nano Lett.* **2015**, *15*, 1660–1666. [CrossRef] [PubMed]

29. Zhao, H.; Wu, J.B.; Zhong, H.X.; Guo, Q.S.; Wang, X.M.; Xia, F.N.; Yang, L.; Tan, P.H.; Wang, H. Interlayer interactions in anisotropic atomically thin rhenium diselenide. *Nano Res.* **2015**, *8*, 3651–3661. [CrossRef]

30. Marzik, J.V.; Kershaw, R.; Dwight, K.; Wold, A. Photoelectronic properties of ReS$_2$ and ReSe$_2$ single crystals. *J. Solid State Chem.* **1984**, *51*, 170–175. [CrossRef]

31. Ridley, B.K. The electron-phonon interaction in quasi-two-dimensional semiconductor quantum-well structures. *J. Phys. C Solid State Phys.* **1982**, *15*, 5899–5917. [CrossRef]

32. Chen, J.H.; Jang, C.; Xiao, S.; Ishigami, M.; Fuhrer, M.S. Intrinsic and extrinsic performance limits of graphene devices on SiO$_2$. *Nat. Nanotechnol.* **2008**, *3*, 206–209. [CrossRef] [PubMed]

33. Fivaz, R.; Mooser, E. Mobility of Charge Carriers in Semiconducting Layer Structures. *Phys. Rev.* **1967**, *163*, 743–755. [CrossRef]

34. Ando, T.; Fowler, A.B.; Stern, F. Electronic properties of two-dimensional systems. *Rev. Mod. Phys.* **1982**, *54*, 437–672. [CrossRef]

35. Hwang, E.H.; Adam, S.; Sarma, S.D. Carrier transport in two-dimensional graphene layers. *Phys. Rev. Lett.* **2007**, *98*, 186806. [CrossRef] [PubMed]

36. Hess, K.; Vogl, P. Remote polar phonon scattering in silicon inversion layers. *Solid State Commun.* **1979**, *30*, 797–799. [CrossRef]

37. Nomura, K.; MacDonald, A.H. Quantum transport of massless Dirac fermions. *Phys. Rev. Lett.* **2007**, *98*, 076602. [CrossRef] [PubMed]

38. Adam, S.; Hwang, E.H.; Sarma, S.D. Scattering mechanisms and Boltzmann transport in graphene. *Phys. E Low-Dimens. Syst. Nanostruct.* **2008**, *40*, 1022–1025. [CrossRef]

39. Paul, T.; Ghatak, S.; Ghosh, A. Percolative switching in transition metal dichalcogenide field-effect transistors at room temperature. *Nanotechnology* **2016**, *27*, 125706. [CrossRef] [PubMed]

40. Podzorov, V.; Gershenson, M.E.; Kloc, C.; Zeis, R.; Bucher, E. High-mobility field-effect transistors based on transition metal dichalcogenides. *Appl. Phys. Lett.* **2004**, *84*, 3301–3303. [CrossRef]

41. Novoselov, K.S.; Jiang, D.; Schedin, F.; Booth, T.J.; Khotkevich, V.V.; Morozov, S.V.; Geim, A.K. Two-dimensional atomic crystals. *Proc. Natl. Acad. Sci. USA* **2005**, *102*, 10451–10453. [CrossRef] [PubMed]

42. Ayari, A.; Cobas, E.; Ogundadegbe, O.; Fuhrer, M.S. Realization and electrical characterization of ultrathin crystals of layered transition-metal dichalcogenides. *J. Appl. Phys.* **2007**, *101*, 014507. [CrossRef]

43. Lee, K.; Kim, H.Y.; Lotya, M.; Coleman, J.N.; Kim, G.T.; Duesberg, G.S. Electrical characteristics of molybdenum disulfide flakes produced by liquid exfoliation. *Adv. Mater.* **2011**, *23*, 4178–4182. [CrossRef] [PubMed]

44. Zhu, W.J.; Perebeinos, V.; Freitag, M.; Avouris, P. Carrier scattering, mobilities, and electrostatic potential in monolayer, bilayer, and trilayer graphene. *Phys. Rev. B* **2009**, *80*, 235402. [CrossRef]

45. Lee, Y.H.; Zhang, X.Q.; Zhang, W.; Chang, M.T.; Lin, C.T.; Chang, K.D.; Yu, Y.C.; Wang, J.T.; Chang, C.S.; Li, L.J.; et al. Synthesis of large-area MoS$_2$ atomic layers with chemical vapor deposition. *Adv. Mater.* **2012**, *24*, 2320–2325. [CrossRef] [PubMed]

46. Kim, S.; Konar, A.; Hwang, W.S.; Lee, J.H.; Lee, J.; Yang, J.; Jung, C.; Kim, H.; Yoo, J.B.; Choi, J.Y.; et al. High-mobility and low-power thin-film transistors based on multilayer MoS$_2$ crystals. *Nat. Commun.* **2012**, *3*, 1011. [CrossRef] [PubMed]

47. Qiu, H.; Pan, L.J.; Yao, Z.N.; Li, J.J.; Shi, Y.; Wang, X.R. Electrical characterization of back-gated bi-layer MoS$_2$ field-effect transistors and the effect of ambient on their performances. *Appl. Phys. Lett.* **2012**, *100*, 123104.

48. Liu, H.; Ye, P.D. MoS$_2$ Dual-Gate MOSFET With Atomic-Layer-Deposited Al2O3 as Top-Gate Dielectric. *IEEE Electron. Device Lett.* **2012**, *33*, 546–548. [CrossRef]

49. Das, S.; Chen, H.Y.; Penumatcha, A.V.; Appenzeller, J. High performance multilayer MoS$_2$ transistors with scandium contacts. *Nano Lett.* **2013**, *13*, 100–105. [CrossRef] [PubMed]

50. Wu, W.; De, D.; Chang, S.C.; Wang, Y.; Peng, H.; Bao, J.; Pei, S.-S. High mobility and high on/off ratio field-effect transistors based on chemical vapor deposited single-crystal MoS$_2$ grains. *Appl. Phys. Lett.* **2013**, *102*, 142106. [CrossRef]

51. Pradhan, N.R.; Rhodes, D.; Zhang, Q.; Talapatra, S.; Terrones, M.; Ajayan, P.M.; Balicas, L. Intrinsic carrier mobility of multi-layered MoS$_2$ field-effect transistors on SiO$_2$. *Appl. Phys. Lett.* **2013**, *102*, 123105. [CrossRef]

52. Bao, W.Z.; Cai, X.H.; Kim, D.; Sridhara, K.; Fuhrer, M.S. High mobility ambipolar MoS$_2$ field-effect transistors: Substrate and dielectric effects. *Appl. Phys. Lett.* **2013**, *102*, 042104. [CrossRef]

53. Desai, S.B.; Madhvapathy, S.R.; Sachid, A.B.; Llinas, J.P.; Wang, Q.; Ahn, G.H.; Pitner, G.; Kim, M.J.; Bokor, J.; Hu, C.; Wong, H.S.P.; Javey, A. MoS$_2$ transistors with 1-nanometer gate lengths. *Science* **2016**, *354*, 99–102. [CrossRef] [PubMed]

54. Larentis, S.; Fallahazad, B.; Tutuc, E. Field-effect transistors and intrinsic mobility in ultra-thin MoSe$_2$ layers. *Appl. Phys. Lett.* **2012**, *101*, 223104. [CrossRef]

55. Chamlagain, B.; Li, Q.; Ghimire, N.J.; Chuang, H.J.; Perera, M.M.; Tu, H.; Xu, Y.; Pan, M.; Xiao, D.; Yan, J.; Mandrus, D.; Zhou, Z. Mobility improvement and temperature dependence in MoSe$_2$ field-effect transistors on parylene-C substrate. *ACS Nano* **2014**, *8*, 5079–5088. [CrossRef] [PubMed]

56. Pradhan, N.R.; Rhodes, D.; Xin, Y.; Memaran, S.; Bhaskaran, L.; Siddiq, M.; Hill, S.; Ajayan, P.M.; Balicas, L. Ambipolar molybdenum diselenide field-effect transistors: Field-effect and Hall mobilities. *ACS Nano* **2014**, *8*, 7923–7929. [CrossRef] [PubMed]

57. Jung, C.; Kim, S.M.; Moon, H.; Han, G.; Kwon, J.; Hong, Y.K.; Omkaram, I.; Yoon, Y.; Kim, S.; Park, J. Highly Crystalline CVD-grown Multilayer MoSe$_2$ Thin Film Transistor for Fast Photodetector. *Sci. Rep.* **2015**, *5*, 15313. [CrossRef] [PubMed]

58. Lin, Y.F.; Xu, Y.; Wang, S.T.; Li, S.L.; Yamamoto, M.; Aparecido-Ferreira, A.; Li, W.; Sun, H.; Nakaharai, S.; Jian, W.B.; Ueno, K.; Tsukagoshi, K. Ambipolar MoTe$_2$ transistors and their applications in logic circuits. *Adv. Mater.* **2014**, *26*, 3263–3269. [CrossRef] [PubMed]

59. Pradhan, N.R.; Rhodes, D.; Feng, S.; Xin, Y.; Memaran, S.; Moon, B.H.; Terrones, H.; Terrones, M.; Balicas, L. Field-Effect Transistors Based on Few-Layered alpha-MoTe$_2$. *ACS Nano* **2014**, *8*, 5911–5920. [CrossRef] [PubMed]

60. Lezama, I.G.; Ubaldini, A.; Longobardi, M.; Giannini, E.; Renner, C.; Kuzmenko, A.B.; Morpurgo, A.F. Surface transport and band gap structure of exfoliated 2H-MoTe$_2$ crystals. *2D Mater.* **2014**, *1*, 021002. [CrossRef]

61. Fathipour, S.; Ma, N.; Hwang, W.S.; Protasenko, V.; Vishwanath, S.; Xing, H.G.; Xu, H.; Jena, D.; Appenzeller, J.; Seabaugh, A. Exfoliated multilayer MoTe$_2$ field-effect transistors. *Appl. Phys. Lett.* **2014**, *105*, 192101. [CrossRef]

62. Xu, H.L.; Fathipour, S.; Kinder, E.W.; Seabaugh, A.C.; Fullerton-Shirey, S.K. Reconfigurable Ion Gating of 2H-MoTe$_2$ Field-Effect Transistors Using Poly(ethylene oxide)-CsClO4 Solid Polymer Electrolyte. *ACS Nano* **2015**, *9*, 4900–4910. [CrossRef] [PubMed]

63. Yin, L.; Zhan, X.Y.; Xu, K.; Wang, F.; Wang, Z.X.; Huang, Y.; Wang, Q.S.; Jiang, C.; He, J. Ultrahigh sensitive MoTe$_2$ phototransistors driven by carrier tunneling. *Appl. Phys. Lett.* **2016**, *108*, 043503. [CrossRef]

64. Octon, T.J.; Nagareddy, V.K.; Russo, S.; Craciun, M.F.; Wright, C.D. Fast High-Responsivity Few-Layer MoTe$_2$ Photodetectors. *Adv. Opt. Mater.* **2016**, *4*, 1750–1754. [CrossRef]

65. Sik Hwang, W.; Remskar, M.; Yan, R.; Protasenko, V.; Tahy, K.; Doo Chae, S.; Zhao, P.; Konar, A.; Xing, H.; Seabaugh, A.; et al. Transistors with chemically synthesized layered semiconductor WS$_2$ exhibiting 105 room temperature modulation and ambipolar behavior. *Appl. Phys. Lett.* **2012**, *101*, 013107. [CrossRef]

66. Braga, D.; Gutierrez Lezama, I.; Berger, H.; Morpurgo, A.F. Quantitative determination of the band gap of WS$_2$ with ambipolar ionic liquid-gated transistors. *Nano Lett.* **2012**, *12*, 5218–5223. [CrossRef] [PubMed]

67. Liu, X.; Hu, J.; Yue, C.; Della Fera, N.; Ling, Y.; Mao, Z.; Wei, J. High performance field-effect transistor based on multilayer tungsten disulfide. *ACS Nano* **2014**, *8*, 10396–10402. [CrossRef] [PubMed]

68. Kumar, J.; Kuroda, M.A.; Bellus, M.Z.; Han, S.J.; Chiu, H.Y. Full-range electrical characteristics of WS$_2$ transistors. *Appl. Phys. Lett.* **2015**, *106*, 123508. [CrossRef]

69. Chuang, H.J.; Tan, X.; Ghimire, N.J.; Perera, M.M.; Chamlagain, B.; Cheng, M.M.; Yan, J.; Mandrus, D.; Tomanek, D.; Zhou, Z. High mobility WSe$_2$ p- and n-type field-effect transistors contacted by highly doped graphene for low-resistance contacts. *Nano Lett.* **2014**, *14*, 3594–3601. [CrossRef] [PubMed]

70. Zhou, H.; Wang, C.; Shaw, J.C.; Cheng, R.; Chen, Y.; Huang, X.; Liu, Y.; Weiss, N.O.; Lin, Z.; Huang, Y.; et al. Large area growth and electrical properties of p-type WSe$_2$ atomic layers. *Nano Lett.* **2015**, *15*, 709–713. [CrossRef] [PubMed]

71. Pradhan, N.R.; Rhodes, D.; Memaran, S.; Poumirol, J.M.; Smirnov, D.; Talapatra, S.; Feng, S.; Perea-Lopez, N.; Elias, A.L.; Terrones, M.; et al. Hall and field-effect mobilities in few layered p-WSe(2) field-effect transistors. *Sci. Rep.* **2015**, *5*, 8979. [CrossRef] [PubMed]

72. Campbell, P.M.; Tarasov, A.; Joiner, C.A.; Tsai, M.Y.; Pavlidis, G.; Graham, S.; Ready, W.J.; Vogel, E.M. Field-effect transistors based on wafer-scale, highly uniform few-layer p-type WSe$_2$. *Nanoscale* **2016**, *8*, 2268–2276. [CrossRef] [PubMed]

73. Liu, B.; Ma, Y.; Zhang, A.; Chen, L.; Abbas, A.N.; Liu, Y.; Shen, C.; Wan, H.; Zhou, C. High-Performance WSe$_2$ Field-Effect Transistors via Controlled Formation of In-Plane Heterojunctions. *ACS Nano* **2016**, *10*, 5153–5160. [CrossRef] [PubMed]

74. Corbet, C.M.; McClellan, C.; Rai, A.; Sonde, S.S.; Tutuc, E.; Banerjee, S.K. Field effect transistors with current saturation and voltage gain in ultrathin ReS$_2$. *ACS Nano* **2015**, *9*, 363–370. [CrossRef] [PubMed]

75. Zhang, E.; Jin, Y.; Yuan, X.; Wang, W.; Zhang, C.; Tang, L.; Liu, S.; Zhou, P.; Hu, W.; Xiu, F. ReS$_2$-Based Field-Effect Transistors and Photodetectors. *Adv. Funct. Mater.* **2015**, *25*, 4076–4082. [CrossRef]

76. Liu, E.; Fu, Y.; Wang, Y.; Feng, Y.; Liu, H.; Wan, X.; Zhou, W.; Wang, B.; Shao, L.; Ho, C.H.; et al. Integrated digital inverters based on two-dimensional anisotropic ReS$_2$ field-effect transistors. *Nat. Commun.* **2015**, *6*, 6991. [CrossRef] [PubMed]

77. Xu, K.; Deng, H.X.; Wang, Z.; Huang, Y.; Wang, F.; Li, S.S.; Luo, J.W.; He, J. Sulfur vacancy activated field effect transistors based on ReS$_2$ nanosheets. *Nanoscale* **2015**, *7*, 15757–15762. [CrossRef] [PubMed]

78. Shim, J.; Oh, A.; Kang, D.H.; Oh, S.; Jang, S.K.; Jeon, J.; Jeon, M.H.; Kim, M.; Choi, C.; Lee, J.; et al. High-Performance 2D Rhenium Disulfide (ReS$_2$) Transistors and Photodetectors by Oxygen Plasma Treatment. *Adv. Mater.* **2016**, *28*, 6985–6992. [CrossRef] [PubMed]

79. Keyshar, K.; Gong, Y.; Ye, G.; Brunetto, G.; Zhou, W.; Cole, D.P.; Hackenberg, K.; He, Y.; Machado, L.; Kabbani, M.; et al. Chemical Vapor Deposition of Monolayer Rhenium Disulfide (ReS$_2$). *Adv. Mater.* **2015**, *27*, 4640–4648. [CrossRef] [PubMed]

80. Corbet, C.M.; Sonde, S.S.; Tutuc, E.; Banerjee, S.K. Improved contact resistance in ReSe$_2$ thin film field-effect transistors. *Appl. Phys. Lett.* **2016**, *108*, 162104. [CrossRef]

81. Hafeez, M.; Gan, L.; Li, H.; Ma, Y.; Zhai, T. Chemical Vapor Deposition Synthesis of Ultrathin Hexagonal ReSe$_2$ Flakes for Anisotropic Raman Property and Optoelectronic Application. *Adv. Mater.* **2016**, *28*, 8296–8301. [CrossRef] [PubMed]

82. Radisavljevic, B.; Radenovic, A.; Brivio, J.; Giacometti, V.; Kis, A. Single-layer MoS$_2$ transistors. *Nat. Nanotechnol.* **2011**, *6*, 147–150. [CrossRef] [PubMed]

83. Ghatak, S.; Pal, A.N.; Ghosh, A. Nature of Electronic States in Atomically Thin MoS$_2$ Field-Effect Transistors. *ACS Nano* **2011**, *5*, 7707–7712. [CrossRef] [PubMed]

84. Van der Zande, A.M.; Huang, P.Y.; Chenet, D.A.; Berkelbach, T.C.; You, Y.; Lee, G.H.; Heinz, T.F.; Reichman, D.R.; Muller, D.A.; Hone, J.C. Grains and grain boundaries in highly crystalline monolayer molybdenum disulphide. *Nat. Mater.* **2013**, *12*, 554–561. [CrossRef] [PubMed]

85. Yazyev, O.V.; Louie, S.G. Electronic transport in polycrystalline graphene. *Nat. Mater.* **2010**, *9*, 806–809. [CrossRef] [PubMed]

86. Hwang, S.W.; Remskar, M.; Yan, R.; Kosel, T.; Park, J.K.; Cho, B.J.; Haensch, W.; Xing, H.; Seabaugh, A.; Jena, D. Comparative study of chemically synthesized and exfoliated multilayer MoS$_2$ field-effect transistors. *Appl. Phys. Lett.* **2013**, *102*, 043116. [CrossRef]

87. Walia, S.; Balendhran, S.; Wang, Y.C.; Ab Kadir, R.; Zoolfakar, A.S.; Atkin, P.; Ou, J.Z.; Sriram, S.; Kalantar-zadeh, K.; Bhaskaran, M. Characterization of metal contacts for two-dimensional MoS$_2$ nanoflakes. *Appl. Phys. Lett.* **2013**, *103*, 232105. [CrossRef]

88. Strait, J.H.; Nene, P.; Rana, F. High intrinsic mobility and ultrafast carrier dynamics in multilayer metal-dichalcogenideMoS$_2$. *Phys. Rev. B* **2014**, *90*, 245402. [CrossRef]

89. Das, S.; Appenzeller, J. Where does the current flow in two-dimensional layered systems? *Nano Lett.* **2013**, *13*, 3396–3402. [CrossRef] [PubMed]

90. Cheng, R.; Jiang, S.; Chen, Y.; Liu, Y.; Weiss, N.; Cheng, H.C.; Wu, H.; Huang, Y.; Duan, X. Few-layer molybdenum disulfide transistors and circuits for high-speed flexible electronics. *Nat. Commun.* **2014**, *5*, 5143. [CrossRef] [PubMed]

91. Wang, H.; Yu, L.; Lee, Y.H.; Fang, W.; Hsu, A.; Herring, P.; Chin, M.; Dubey, M.; Li, L.J.; Kong, J.; et al. Large-scale 2D electronics based on single-layer MoS$_2$ grown by chemical vapor deposition. *IEDM Tech. Dig.* **2012**, 4.6.1–4.6.4. [CrossRef]

92. Das, S.R.; Kwon, J.; Prakash, A.; Delker, C.J.; Das, S.; Janes, D.B. Low-frequency noise in MoSe$_2$ field effect transistors. *Appl. Phys. Lett.* **2015**, *106*, 083507. [CrossRef]

93. Kwon, H.-J.; Kang, H.; Jang, J.; Kim, S.; Grigoropoulos, C.P. Analysis of flicker noise in two-dimensional multilayer MoS$_2$ transistors. *Appl. Phys. Lett.* **2014**, *104*, 083110. [CrossRef]

94. Na, J.; Joo, M.K.; Shin, M.; Huh, J.; Kim, J.S.; Piao, M.; Jin, J.E.; Jang, H.K.; Choi, H.J.; Shim, J.H.; et al. Low-frequency noise in multilayer MoS$_2$ field-effect transistors: the effect of high-k passivation. *Nanoscale* **2014**, *6*, 433–441. [CrossRef] [PubMed]

95. Liu, L.; Kumar, S.B.; Ouyang, Y.; Guo, J. Performance Limits of Monolayer Transition Metal Dichalcogenide Transistors. *IEEE Trans. Electron. Devices* **2011**, *58*, 3042–3047. [CrossRef]

96. Yoon, Y.; Ganapathi, K.; Salahuddin, S. How good can monolayer MoS$_2$ transistors be? *Nano Lett.* **2011**, *11*, 3768–3773. [CrossRef] [PubMed]

97. Radisavljevic, B.; Whitwick, M.B.; Kis, A. Integrated Circuits and Logic Operations Based on Single-Layer MoS$_2$. *ACS Nano* **2011**, *5*, 9934–9938. [CrossRef] [PubMed]

98. Zhan, Y.; Liu, Z.; Najmaei, S.; Ajayan, P.M.; Lou, J. Large-area vapor-phase growth and characterization of MoS$_2$ atomic layers on a SiO$_2$ substrate. *Small* **2012**, *8*, 966–971. [CrossRef] [PubMed]

99. Late, D.J.; Liu, B.; Matte, H.S.; Dravid, V.P.; Rao, C.N. Hysteresis in single-layer MoS$_2$ field effect transistors. *ACS Nano* **2012**, *6*, 5635–5641. [CrossRef] [PubMed]

100. Lin, M.-W.; Liu, L.Z.; Lan, Q.; Tan, X.B.; Dhindsa, K.S.; Zeng, P.; Naik, V.M.; Cheng, M.M.C.; Zhou, Z.X. Mobility enhancement and highly efficient gating of monolayer MoS$_2$transistors with polymer electrolyte. *J. Phys. D Appl. Phys.* **2012**, *45*, 345102. [CrossRef]

101. Lee, H.S.; Min, S.W.; Park, M.K.; Lee, Y.T.; Jeon, P.J.; Kim, J.H.; Ryu, S.; Im, S. MoS$_2$ nanosheets for top-gate nonvolatile memory transistor channel. *Small* **2012**, *8*, 3111–3115. [CrossRef] [PubMed]

102. Radisavljevic, B.; Whitwick, M.B.; Kis, A. Small-signal amplifier based on single-layer MoS$_2$. *Appl. Phys. Lett.* **2012**, *101*, 043103. [CrossRef]

103. Lembke, D.; Kis, A. Breakdown of high-performance monolayer MoS$_2$ transistors. *ACS Nano* **2012**, *6*, 10070–10075. [CrossRef] [PubMed]

104. Baugher, B.W.; Churchill, H.O.; Yang, Y.; Jarillo-Herrero, P. Intrinsic electronic transport properties of high-quality monolayer and bilayer MoS$_2$. *Nano Lett.* **2013**, *13*, 4212–4216. [CrossRef] [PubMed]

105. Park, W.; Baik, J.; Kim, T.Y.; Cho, K.; Hong, W.K.; Shin, H.J.; Lee, T. Photoelectron spectroscopic imaging and device applications of large-area patternable single-layer MoS$_2$ synthesized by chemical vapor deposition. *ACS Nano* **2014**, *8*, 4961–4968. [CrossRef] [PubMed]

106. Sanne, A.; Ghosh, R.; Rai, A.; Movva, H.C.P.; Sharma, A.; Rao, R.; Mathew, L.; Banerjee, S.K. Top-gated chemical vapor deposited MoS$_2$ field-effect transistors on Si3N4 substrates. *Appl. Phys. Lett.* **2015**, *106*, 062101. [CrossRef]

107. Shao, P.-Z.; Zhao, H.M.; Cao, H.W.; Wang, X.F.; Pang, Y.; Li, Y.X.; Deng, N.Q.; Zhang, J.; Zhang, G.Y.; Yang, Y.; et al. Enhancement of carrier mobility in MoS$_2$ field effect transistors by a SiO$_2$ protective layer. *Appl. Phys. Lett.* **2016**, *108*, 203105. [CrossRef]

108. Amani, M.; Chin, M.L.; Birdwell, A.G.; O'Regan, T.P.; Najmaei, S.; Liu, Z.; Ajayan, P.M.; Lou, J.; Dubey, M. Electrical performance of monolayer MoS$_2$ field-effect transistors prepared by chemical vapor deposition. *Appl. Phys. Lett.* **2013**, *102*, 193107. [CrossRef]

109. Lu, X.; Utama, M.I.; Lin, J.; Gong, X.; Zhang, J.; Zhao, Y.; Pantelides, S.T.; Wang, J.; Dong, Z.; Liu, Z.; et al. Large-area synthesis of monolayer and few-layer MoSe$_2$ films on SiO$_2$ substrates. *Nano Lett.* **2014**, *14*, 2419–2425. [CrossRef] [PubMed]

110. Wang, X.; Gong, Y.; Shi, G.; Chow, W.L.; Keyshar, K.; Ye, G.; Vajtai, R.; Lou, J.; Liu, Z.; Ringe, E.; et al. Chemical vapor deposition growth of crystalline monolayer MoSe$_2$. *ACS Nano* **2014**, *8*, 5125–5131. [CrossRef] [PubMed]

111. Chang, Y.H.; Zhang, W.; Zhu, Y.; Han, Y.; Pu, J.; Chang, J.K.; Hsu, W.T.; Huang, J.K.; Hsu, C.L.; Chiu, M.H.; et al. Monolayer MoSe$_2$ grown by chemical vapor deposition for fast photodetection. *ACS Nano* **2014**, *8*, 8582–8590. [CrossRef] [PubMed]

112. Ovchinnikov, D.; Allain, A.; Huang, Y.S.; Dumcenco, D.; Kis, A. Electrical transport properties of single-layer WS$_2$. *ACS Nano* **2014**, *8*, 8174–8181. [CrossRef] [PubMed]

113. Xu, Z.Q.; Zhang, Y.; Lin, S.; Zheng, C.; Zhong, Y.L.; Xia, X.; Li, Z.; Sophia, P.J.; Fuhrer, M.S.; Cheng, Y.B.; et al. Synthesis and Transfer of Large-Area Monolayer WS$_2$ Crystals: Moving Toward the Recyclable Use of Sapphire Substrates. *ACS Nano* **2015**, *9*, 6178–6187. [CrossRef] [PubMed]

114. Hanbicki, A.T.; Currie, M.; Kioseoglou, G.; Friedman, A.L.; Jonker, B.T. Measurement of high exciton binding energy in the monolayer transition-metal dichalcogenides WS$_2$ and WSe$_2$. *Solid State Commun.* **2015**, *203*, 16–20. [CrossRef]

115. Fang, H.; Chuang, S.; Chang, T.C.; Takei, K.; Takahashi, T.; Javey, A. High-performance single layered WSe(2) p-FETs with chemically doped contacts. *Nano Lett.* **2012**, *12*, 3788–3792. [CrossRef] [PubMed]

116. Huang, J.K.; Pu, J.; Hsu, C.L.; Chiu, M.H.; Juang, Z.Y.; Chang, Y.H.; Chang, W.H.; Iwasa, Y.; Takenobu, T.; Li, L.J. Large-Area Synthesis of Highly Crystalline WSe$_2$ Monolayers and Device Applications. *ACS Nano* **2014**, *8*, 923–930. [CrossRef] [PubMed]

117. Allain, A.; Kis, A. Electron and hole mobilities in single-layer WSe$_2$. *ACS Nano* **2014**, *8*, 7180–7185. [CrossRef] [PubMed]

118. Kawamura, T.; Das Sarma, S. Temperature dependence of the low-temperature mobility in ultrapureAlxGa1−xAs/GaAs heterojunctions: Acoustic-phonon scattering. *Phys. Rev. B* **1990**, *42*, 3725–3728. [CrossRef]

119. Hwang, E.H.; Das Sarma, S. Limit to two-dimensional mobility in modulation-doped GaAs quantum structures: How to achieve a mobility of 100 million. *Phys. Rev. B* **2008**, *77*, 235437. [CrossRef]

120. Kaasbjerg, K.; Thygesen, K.S.; Jacobsen, K.W. Phonon-limited mobility in n-type single-layer MoS$_2$ from first principles. *Phys. Rev. B* **2012**, *85*, 115317. [CrossRef]

121. Zhang, W.X.; Huang, Z.S.; Zhang, W.L.; Li, Y.R. Two-dimensional semiconductors with possible high room temperature mobility. *Nano Res.* **2014**, *7*, 1731–1737. [CrossRef]

122. Jena, D.; Konar, A. Enhancement of carrier mobility in semiconductor nanostructures by dielectric engineering. *Phys. Rev. Lett.* **2007**, *98*, 136805. [CrossRef] [PubMed]

123. Chen, F.; Xia, J.; Ferry, D.K.; Tao, N. Dielectric screening enhanced performance in graphene FET. *Nano Lett.* **2009**, *9*, 2571–2574. [CrossRef] [PubMed]

124. Konar, A.; Fang, T.; Jena, D. Effect of high-κ gate dielectrics on charge transport in graphene-based field effect transistors. *Phys. Rev. B* **2010**, *82*, 115452. [CrossRef]

125. Newaz, A.K.; Puzyrev, Y.S.; Wang, B.; Pantelides, S.T.; Bolotin, K.I. Probing charge scattering mechanisms in suspended graphene by varying its dielectric environment. *Nat. Commun.* **2012**, *3*, 734. [CrossRef] [PubMed]

126. Radisavljevic, B.; Kis, A. Mobility engineering and a metal-insulator transition in monolayer MoS(2). *Nat. Mater.* **2013**, *12*, 815–820. [CrossRef] [PubMed]

127. Sakaki, H.; Noda, T.; Hirakawa, K.; Tanaka, M.; Matsusue, T. Interface roughness scattering in GaAs/AlAs quantum wells. *Appl. Phys. Lett.* **1987**, *51*, 1934–1936. [CrossRef]

128. Popov, I.; Seifert, G.; Tomanek, D. Designing electrical contacts to MoS$_2$ monolayers: A computational study. *Phys. Rev. Lett.* **2012**, *108*, 156802. [CrossRef] [PubMed]

129. Kang, J.H.; Liu, W.; Sarkar, D.; Jena, D.; Banerjee, K. Computational Study of Metal Contacts to Monolayer Transition-Metal Dichalcogenide Semiconductors. *Phys. Rev. X* **2014**, *4*, 031005. [CrossRef]

130. Wang, Y.; Yang, R.X.; Quhe, R.; Zhong, H.; Cong, L.; Ye, M.; Ni, Z.; Song, Z.; Yang, J.; Shi, J.; et al. Does p-type ohmic contact exist in WSe$_2$-metal interfaces? *Nanoscale* **2016**, *8*, 1179–1191. [CrossRef] [PubMed]

131. Zhong, H.; Quhe, R.; Wang, Y.; Ni, Z.; Ye, M.; Song, Z.; Pan, Y.; Yang, J.; Yang, L.; Lei, M.; et al. Interfacial Properties of Monolayer and Bilayer MoS$_2$ Contacts with Metals: Beyond the Energy Band Calculations. *Sci. Rep.* **2016**, *6*, 21786. [CrossRef] [PubMed]

132. Gong, C.; Huang, C.; Miller, J.; Cheng, L.; Hao, Y.; Cobden, D.; Kim, J.; Ruoff, R.S.; Wallace, R.M.; Cho, K.; et al. Metal contacts on physical vapor deposited monolayer MoS$_2$. *ACS Nano* **2013**, *7*, 11350–11357. [CrossRef] [PubMed]

133. Leong, W.S.; Luo, X.; Li, Y.; Khoo, K.H.; Quek, S.Y.; Thong, J.T. Low resistance metal contacts to MoS$_2$ devices with nickel-etched-graphene electrodes. *ACS Nano* **2015**, *9*, 869–877. [CrossRef] [PubMed]

134. Farmanbar, M.; Brocks, G. Controlling the Schottky barrier at MoS$_2$/metal contacts by inserting a BN monolayer. *Phys. Rev. B* **2015**, *91*, 161304. [CrossRef]

135. Chuang, H.J.; Chamlagain, B.; Koehler, M.; Perera, M.M.; Yan, J.; Mandrus, D.; Tomanek, D.; Zhou, Z. Low-Resistance 2D/2D Ohmic Contacts: A Universal Approach to High-Performance WSe$_2$, MoS$_2$, and MoSe$_2$ Transistors. *Nano Lett.* **2016**, *16*, 1896–1902. [CrossRef] [PubMed]

136. Farmanbar, M.; Brocks, G. Ohmic Contacts to 2D Semiconductors through van der Waals Bonding. *Adv. Electron. Mater.* **2016**, *2*, 1500405. [CrossRef]

137. Chen, J.R.; Odenthal, P.M.; Swartz, A.G.; Floyd, G.C.; Wen, H.; Luo, K.Y.; Kawakami, R.K. Control of Schottky barriers in single layer MoS$_2$ transistors with ferromagnetic contacts. *Nano Lett.* **2013**, *13*, 3106–3110. [CrossRef] [PubMed]

138. Dankert, A.; Langouche, L.; Kamalakar, M.V.; Dash, S.P. High-performance molybdenum disulfide field-effect transistors with spin tunnel contacts. *ACS Nano* **2014**, *8*, 476–482. [CrossRef] [PubMed]

139. Kappera, R.; Voiry, D.; Yalcin, S.E.; Branch, B.; Gupta, G.; Mohite, A.D.; Chhowalla, M. Phase-engineered low-resistance contacts for ultrathin MoS$_2$ transistors. *Nat. Mater.* **2014**, *13*, 1128–1134. [CrossRef] [PubMed]

140. Kappera, R.; Voiry, D.; Yalcin, S.E.; Jen, W.; Acerce, M.; Torrel, S.; Branch, B.; Lei, S.D.; Chen, W.B.; Najmaei, S.; et al. Metallic 1T phase source/drain electrodes for field effect transistors from chemical vapor deposited MoS$_2$. *APL Mater.* **2014**, *2*, 092516. [CrossRef]

141. Cho, S.; Kim, S.; Kim, J.H.; Zhao, J.; Seok, J.; Keum, D.H.; Baik, J.; Choe, D.H.; Chang, K.J.; Suenaga, K.; et al. DEVICE TECHNOLOGY. Phase patterning for ohmic homojunction contact in MoTe$_2$. *Science* **2015**, *349*, 625–628. [CrossRef] [PubMed]

142. Ma, Y.; Liu, B.; Zhang, A.; Chen, L.; Fathi, M.; Shen, C.; Abbas, A.N.; Ge, M.; Mecklenburg, M.; Zhou, C. Reversible Semiconducting-to-Metallic Phase Transition in Chemical Vapor Deposition Grown Monolayer WSe$_2$ and Applications for Devices. *ACS Nano* **2015**, *9*, 7383–7391. [CrossRef] [PubMed]

143. Sangwan, V.K.; Arnold, H.N.; Jariwala, D.; Marks, T.J.; Lauhon, L.J.; Hersam, M.C. Low-frequency electronic noise in single-layer MoS$_2$ transistors. *Nano Lett.* **2013**, *13*, 4351–4355. [CrossRef] [PubMed]

144. Xie, X.; Sarkar, D.; Liu, W.; Kang, J.; Marinov, O.; Deen, M.J.; Banerjee, K. Low-frequency noise in bilayer MoS$_2$ transistor. *ACS Nano* **2014**, *8*, 5633–5640. [CrossRef] [PubMed]

145. Sharma, D.; Amani, M.; Motayed, A.; Shah, P.B.; Birdwell, A.G.; Najmaei, S.; Ajayan, P.M.; Lou, J.; Dubey, M.; Li, Q.; et al. Electrical transport and low-frequency noise in chemical vapor deposited single-layer MoS$_2$ devices. *Nanotechnology* **2014**, *25*, 155702. [CrossRef] [PubMed]

146. Ghatak, S.; Mukherjee, S.; Jain, M.; Sarma, D.D.; Ghosh, A. Microscopic origin of low frequency noise in MoS$_2$ field-effect transistors. *APL Mater.* **2014**, *2*, 092515. [CrossRef]

147. Lin, Y.F.; Xu, Y.; Lin, C.Y.; Suen, Y.W.; Yamamoto, M.; Nakaharai, S.; Ueno, K.; Tsukagoshi, K. Origin of Noise in Layered MoTe(2) Transistors and its Possible Use for Environmental Sensors. *Adv. Mater.* **2015**, *27*, 6612–6619. [CrossRef] [PubMed]

148. Renteria, J.; Samnakay, R.; Rumyantsev, S.L.; Jiang, C.; Goli, P.; Shur, M.S.; Balandin, A.A. Low-frequency 1/f noise in MoS$_2$ transistors: Relative contributions of the channel and contacts. *Appl. Phys. Lett.* **2014**, *104*, 153104. [CrossRef]

149. Zhang, M.; Wu, J.; Zhu, Y.; Dumcenco, D.O.; Hong, J.; Mao, N.; Deng, S.; Chen, Y.; Yang, Y.; Jin, C.; et al. Two-dimensional molybdenum tungsten diselenide alloys: Photoluminescence, Raman scattering, and electrical transport. *ACS Nano* **2014**, *8*, 7130–7137. [CrossRef] [PubMed]

150. Zhang, Y.; Ye, J.; Matsuhashi, Y.; Iwasa, Y. Ambipolar MoS$_2$ thin flake transistors. *Nano Lett.* **2012**, *12*, 1136–1140. [CrossRef] [PubMed]

151. Zhang, Y.J.; Ye, J.T.; Yomogida, Y.; Takenobu, T.; Iwasa, Y. Formation of a stable p-n junction in a liquid-gated MoS$_2$ ambipolar transistor. *Nano Lett.* **2013**, *13*, 3023–3028. [CrossRef] [PubMed]

152. Shi, W.; Ye, J.; Zhang, Y.; Suzuki, R.; Yoshida, M.; Miyazaki, J.; Inoue, N.; Saito, Y.; Iwasa, Y. Superconductivity Series in Transition Metal Dichalcogenides by Ionic Gating. *Sci. Rep.* **2015**, *5*, 12534. [CrossRef] [PubMed]

153. Groenendijk, D.J.; Buscema, M.; Steele, G.A.; Michaelis de Vasconcellos, S.; Bratschitsch, R.; van der Zant, H.S.; Castellanos-Gomez, A. Photovoltaic and photothermoelectric effect in a double-gated WSe$_2$ device. *Nano Lett.* **2014**, *14*, 5846–5852. [CrossRef] [PubMed]

154. Fontana, M.; Deppe, T.; Boyd, A.K.; Rinzan, M.; Liu, A.Y.; Paranjape, M.; Barbara, P. Electron-hole transport and photovoltaic effect in gated MoS$_2$ Schottky junctions. *Sci. Rep.* **2013**, *3*, 1634. [CrossRef] [PubMed]

155. Das, S.; Appenzeller, J. WSe$_2$ field effect transistors with enhanced ambipolar characteristics. *Appl. Phys. Lett.* **2013**, *103*, 103501. [CrossRef]

156. Fang, H.; Tosun, M.; Seol, G.; Chang, T.C.; Takei, K.; Guo, J.; Javey, A. Degenerate n-doping of few-layer transition metal dichalcogenides by potassium. *Nano Lett.* **2013**, *13*, 1991–1995. [CrossRef] [PubMed]

157. Pospischil, A.; Furchi, M.M.; Mueller, T. Solar-energy conversion and light emission in an atomic monolayer p-n diode. *Nat. Nanotechnol.* **2014**, *9*, 257–261. [CrossRef] [PubMed]

158. Baugher, B.W.; Churchill, H.O.; Yang, Y.; Jarillo-Herrero, P. Optoelectronic devices based on electrically tunable p-n diodes in a monolayer dichalcogenide. *Nat. Nanotechnol.* **2014**, *9*, 262–267. [CrossRef] [PubMed]

159. Ross, J.S.; Klement, P.; Jones, A.M.; Ghimire, N.J.; Yan, J.; Mandrus, D.G.; Taniguchi, T.; Watanabe, K.; Kitamura, K.; Yao, W.; et al. Electrically tunable excitonic light-emitting diodes based on monolayer WSe$_2$ p-n junctions. *Nat. Nanotechnol.* **2014**, *9*, 268–272. [CrossRef] [PubMed]

160. Nakaharai, S.; Yamamoto, M.; Ueno, K.; Lin, Y.F.; Li, S.L.; Tsukagoshi, K. Electrostatically Reversible Polarity of Ambipolar alpha-MoTe$_2$ Transistors. *ACS Nano* **2015**, *9*, 5976–5983. [CrossRef] [PubMed]

161. Geim, A.K.; Grigorieva, I.V. Van der Waals heterostructures. *Nature* **2013**, *499*, 419–425. [CrossRef] [PubMed]

162. Novoselov, K.S.; Mishchenko, A.; Carvalho, A.; Castro Neto, A.H. 2D materials and van der Waals heterostructures. *Science* **2016**, *353*, aac9439. [CrossRef] [PubMed]

163. Yang, H.; Heo, J.; Park, S.; Song, H.J.; Seo, D.H.; Byun, K.E.; Kim, P.; Yoo, I.; Chung, H.J.; Kim, K. Graphene barristor, a triode device with a gate-controlled Schottky barrier. *Science* **2012**, *336*, 1140–1143. [CrossRef] [PubMed]

164. Britnell, L.; Gorbachev, R.V.; Jalil, R.; Belle, B.D.; Schedin, F.; Mishchenko, A.; Georgiou, T.; Katsnelson, M.I.; Eaves, L.; Morozov, S.V.; et al. Field-effect tunneling transistor based on vertical graphene heterostructures. *Science* **2012**, *335*, 947–950. [CrossRef] [PubMed]

165. Yu, W.J.; Li, Z.; Zhou, H.; Chen, Y.; Wang, Y.; Huang, Y.; Duan, X. Vertically stacked multi-heterostructures of layered materials for logic transistors and complementary inverters. *Nat. Mater.* **2013**, *12*, 246–252. [CrossRef] [PubMed]

166. Georgiou, T.; Jalil, R.; Belle, B.D.; Britnell, L.; Gorbachev, R.V.; Morozov, S.V.; Kim, Y.J.; Gholinia, A.; Haigh, S.J.; Makarovsky, O.; et al. Vertical field-effect transistor based on graphene-WS$_2$ heterostructures for flexible and transparent electronics. *Nat. Nanotechnol.* **2013**, *8*, 100–103. [CrossRef] [PubMed]

167. Sata, Y.; Moriya, R.; Morikawa, S.; Yabuki, N.; Masubuchi, S.; Machida, T. Electric field modulation of Schottky barrier height in graphene/MoSe$_2$ van der Waals heterointerface. *Appl. Phys. Lett.* **2015**, *107*, 023109. [CrossRef]

168. Choi, Y.; Kang, J.; Jariwala, D.; Kang, M.S.; Marks, T.J.; Hersam, M.C.; Cho, J.H. Low-Voltage Complementary Electronics from Ion-Gel-Gated Vertical Van der Waals Heterostructures. *Adv. Mater.* **2016**, *28*, 3742–3748. [CrossRef] [PubMed]

169. Shih, C.J.; Wang, Q.H.; Son, Y.; Jin, Z.; Blankschtein, D.; Strano, M.S. Tuning On-Off Current Ratio and Field-Effect Mobility in a MoS$_2$-Graphene Heterostructure via Schottky Barrier Modulation. *ACS Nano* **2014**, *8*, 5790–5798. [CrossRef] [PubMed]

170. Kwak, J.Y.; Hwang, J.; Calderon, B.; Alsalman, H.; Munoz, N.; Schutter, B.; Spencer, M.G. Electrical characteristics of multilayer MoS$_2$ FET's with MoS$_2$/graphene heterojunction contacts. *Nano Lett.* **2014**, *14*, 4511–4516. [CrossRef] [PubMed]

171. Moriya, R.; Yamaguchi, T.; Inoue, Y.; Morikawa, S.; Sata, Y.; Masubuchi, S.; Machida, T. Large current modulation in exfoliated-graphene/MoS$_2$/metal vertical heterostructures. *Appl. Phys. Lett.* **2014**, *105*, 083119. [CrossRef]

172. Lin, Y.F.; Li, W.; Li, S.L.; Xu, Y.; Aparecido-Ferreira, A.; Komatsu, K.; Sun, H.; Nakaharai, S.; Tsukagoshi, K. Barrier inhomogeneities at vertically stacked graphene-based heterostructures. *Nanoscale* **2014**, *6*, 795–799. [CrossRef] [PubMed]

173. Qiu, D.; Kim, E.K. Electrically Tunable and Negative Schottky Barriers in Multi-layered Graphene/MoS$_2$ Heterostructured Transistors. *Sci. Rep.* **2015**, *5*, 13743. [CrossRef] [PubMed]

174. Cheng, H.C.; Wang, G.; Li, D.; He, Q.; Yin, A.; Liu, Y.; Wu, H.; Ding, M.; Huang, Y.; Duan, X. van der Waals Heterojunction Devices Based on Organohalide Perovskites and Two-Dimensional Materials. *Nano Lett.* **2016**, *16*, 367–373. [CrossRef] [PubMed]

175. Shim, J.; Kim, H.S.; Shim, Y.S.; Kang, D.H.; Park, H.Y.; Lee, J.; Jeon, J.; Jung, S.J.; Song, Y.J.; Jung, W.S.; et al. Extremely Large Gate Modulation in Vertical Graphene/WSe$_2$ Heterojunction Barristor Based on a Novel Transport Mechanism. *Adv. Mater.* **2016**, *28*, 5293–5299. [CrossRef] [PubMed]

176. Bernardi, M.; Palummo, M.; Grossman, J.C. Extraordinary sunlight absorption and one nanometer thick photovoltaics using two-dimensional monolayer materials. *Nano Lett.* **2013**, *13*, 3664–3670. [CrossRef] [PubMed]

177. Britnell, L.; Ribeiro, R.M.; Eckmann, A.; Jalil, R.; Belle, B.D.; Mishchenko, A.; Kim, Y.J.; Gorbachev, R.V.; Georgiou, T.; Morozov, S.V.; et al. Strong light-matter interactions in heterostructures of atomically thin films. *Science* **2013**, *340*, 1311–1314. [CrossRef] [PubMed]

178. Eda, G.; Maier, S.A. Two-dimensional crystals: Managing light for optoelectronics. *ACS Nano* **2013**, *7*, 5660–5665. [CrossRef] [PubMed]

179. Yu, W.J.; Liu, Y.; Zhou, H.; Yin, A.; Li, Z.; Huang, Y.; Duan, X. Highly efficient gate-tunable photocurrent generation in vertical heterostructures of layered materials. *Nat. Nanotechnol.* **2013**, *8*, 952–958. [CrossRef] [PubMed]

180. Massicotte, M.; Schmidt, P.; Vialla, F.; Schadler, K.G.; Reserbat-Plantey, A.; Watanabe, K.; Taniguchi, T.; Tielrooij, K.J.; Koppens, F.H. Picosecond photoresponse in van der Waals heterostructures. *Nat. Nanotechnol.* **2016**, *11*, 42–46. [CrossRef] [PubMed]

181. Long, M.; Liu, E.; Wang, P.; Gao, A.; Xia, H.; Luo, W.; Wang, B.; Zeng, J.; Fu, Y.; Xu, K.; et al. Broadband Photovoltaic Detectors Based on an Atomically Thin Heterostructure. *Nano Lett.* **2016**, *16*, 2254–2259. [CrossRef] [PubMed]

182. Furchi, M.M.; Pospischil, A.; Libisch, F.; Burgdorfer, J.; Mueller, T. Photovoltaic effect in an electrically tunable van der Waals heterojunction. *Nano Lett.* **2014**, *14*, 4785–4791. [CrossRef] [PubMed]

183. Lee, C.H.; Lee, G.H.; van der Zande, A.M.; Chen, W.; Li, Y.; Han, M.; Cui, X.; Arefe, G.; Nuckolls, C.; Heinz, T.F.; et al. Atomically thin p-n junctions with van der Waals heterointerfaces. *Nat. Nanotechnol.* **2014**, *9*, 676–681. [CrossRef] [PubMed]

184. Cheng, R.; Li, D.; Zhou, H.; Wang, C.; Yin, A.; Jiang, S.; Liu, Y.; Chen, Y.; Huang, Y.; Duan, X. Electroluminescence and photocurrent generation from atomically sharp WSe$_2$/MoS$_2$ heterojunction p-n diodes. *Nano Lett.* **2014**, *14*, 5590–5597. [CrossRef] [PubMed]

185. Deng, Y.X.; Luo, Z.; Conrad, N.J.; Liu, H.; Gong, Y.J.; Najmaei, S.; Ajayan, P.M.; Lou, J.; Xu, X.F.; Ye, P.D. Black Phosphorus-Monolayer MoS$_2$ van der Waals Heterojunction p–n Diode. *ACS Nano* **2014**, *8*, 8292–8299. [CrossRef] [PubMed]

186. Wang, Q.H.; Kalantar-Zadeh, K.; Kis, A.; Coleman, J.N.; Strano, M.S. Electronics and optoelectronics of two-dimensional transition metal dichalcogenides. *Nat. Nanotechnol.* **2012**, *7*, 699–712. [CrossRef] [PubMed]

187. Bertolazzi, S.; Brivio, J.; Kis, A. Stretching and Breaking of Ultrathin MoS$_2$. *Acs Nano* **2011**, *5*, 9703–9709. [CrossRef] [PubMed]

188. Jariwala, D.; Sangwan, V.K.; Lauhon, L.J.; Marks, T.J.; Hersam, M.C. Emerging device applications for semiconducting two-dimensional transition metal dichalcogenides. *ACS Nano* **2014**, *8*, 1102–1120. [CrossRef] [PubMed]

189. Lopez-Sanchez, O.; Alarcon Llado, E.; Koman, V.; Fontcuberta i Morral, A.; Radenovic, A.; Kis, A. Light generation and harvesting in a van der Waals heterostructure. *ACS Nano* **2014**, *8*, 3042–3048. [CrossRef] [PubMed]

190. Esmaeili-Rad, M.R.; Salahuddin, S. High performance molybdenum disulfide amorphous silicon heterojunction photodetector. *Sci. Rep.* **2013**, *3*, 2345. [CrossRef] [PubMed]

191. Gourmelon, E.; Lignier, O.; Hadouda, H.; Couturier, G.; Bernwde, J.C.; Tedd, J.; Pouzet, J.; Salardenne, J. MS$_2$ (M = W, Mo) Photosensitive thin film for solar cells. *Sol. Energy Mater. Sol. Cells* **1997**, *46*, 115–121. [CrossRef]

192. Thomalla, M.; Tributsch, H. Photosensitization of nanostructured TiO$_2$ with WS$_2$ quantum sheets. *J. Phys. Chem. B* **2006**, *110*, 12167–12171. [CrossRef] [PubMed]

193. Shanmugam, M.; Bansal, T.; Durcan, C.A.; Yu, B. Molybdenum disulphide/titanium dioxide nanocomposite-poly 3-hexylthiophene bulk heterojunction solar cell. *Appl. Phys. Lett.* **2012**, *100*, 153901. [CrossRef]

194. Shanmugam, M.; Durcan, C.A.; Yu, B. Layered semiconductor molybdenum disulfide nanomembrane based Schottky-barrier solar cells. *Nanoscale* **2012**, *4*, 7399–7405. [CrossRef] [PubMed]

195. Wi, S.; Kim, H.; Chen, M.; Nam, H.; Guo, L.J.; Meyhofer, E.; Liang, X. Enhancement of photovoltaic response in multilayer MoS$_2$ induced by plasma doping. *ACS Nano* **2014**, *8*, 5270–5281. [CrossRef] [PubMed]

196. Polman, A.; Atwater, H.A. Photonic design principles for ultrahigh-efficiency photovoltaics. *Nat. Mater.* **2012**, *11*, 174–177. [CrossRef] [PubMed]

197. Lin, J.D.; Li, H.; Zhang, H.; Chen, W. Plasmonic enhancement of photocurrent in MoS$_2$ field-effect-transistor. *Appl. Phys. Lett.* **2013**, *102*, 203109. [CrossRef]

198. Lee, H.S.; Min, S.W.; Chang, Y.G.; Park, M.K.; Nam, T.; Kim, H.; Kim, J.H.; Ryu, S.; Im, S. MoS$_2$ nanosheet phototransistors with thickness-modulated optical energy gap. *Nano Lett.* **2012**, *12*, 3695–3700. [CrossRef] [PubMed]

199. Feng, J.; Qian, X.F.; Huang, C.W.; Li, J. Strain-engineered artificial atom as a broad-spectrum solar energy funnel. *Nat. Photonics* **2012**, *6*, 865–871. [CrossRef]

200. Carladous, A.; Coratger, R.; Ajustron, F.; Seine, G.; Péchou, R.; Beauvillain, J. Light emission from spectral analysis of Au/MoS$_2$ nanocontacts stimulated by scanning tunneling microscopy. *Phys. Rev. B* **2002**, *66*, 045401. [CrossRef]

201. Kirmayer, S.; Aharon, E.; Dovgolevsky, E.; Kalina, M.; Frey, G.L. Self-assembled lamellar MoS$_2$, SnS$_2$ and SiO$_2$ semiconducting polymer nanocomposites. *Philos. Trans. A Math. Phys. Eng. Sci.* **2007**, *365*, 1489–1508. [CrossRef] [PubMed]

202. Sundaram, R.S.; Engel, M.; Lombardo, A.; Krupke, R.; Ferrari, A.C.; Avouris, P.; Steiner, M. Electroluminescence in single layer MoS$_2$. *Nano Lett.* **2013**, *13*, 1416–1421. [CrossRef] [PubMed]

203. Zhang, Y.J.; Oka, T.; Suzuki, R.; Ye, J.T.; Iwasa, Y. Electrically Switchable Chiral Light-Emitting Transistor. *Science* **2014**, *344*, 725–728. [CrossRef] [PubMed]

204. Withers, F.; Del Pozo-Zamudio, O.; Mishchenko, A.; Rooney, A.P.; Gholinia, A.; Watanabe, K.; Taniguchi, T.; Haigh, S.J.; Geim, A.K.; Tartakovskii, A.I.; et al. Light-emitting diodes by band-structure engineering in van der Waals heterostructures. *Nat. Mater.* **2015**, *14*, 301–306. [CrossRef] [PubMed]

205. Withers, F.; Del Pozo-Zamudio, O.; Schwarz, S.; Dufferwiel, S.; Walker, P.M.; Godde, T.; Rooney, A.P.; Gholinia, A.; Woods, C.R.; Blake, P.; et al. WSe(2) Light-Emitting Tunneling Transistors with Enhanced Brightness at Room Temperature. *Nano Lett.* **2015**, *15*, 8223–8228. [CrossRef] [PubMed]

206. Muccini, M.; Toffanin, S. *Organic Light-Emitting Transistors, in Organic Light-Emitting Transistors*; John Wiley & Sons, Inc.: Hoboken, NJ, USA, 2016; pp. 45–85.

207. Clark, G.; Schaibley, J.R.; Ross, J.; Taniguchi, T.; Watanabe, K.; Hendrickson, J.R.; Mou, S.; Yao, W.; Xu, X. Single Defect Light-Emitting Diode in a van der Waals Heterostructure. *Nano Lett.* **2016**, *16*, 3944–3948. [CrossRef] [PubMed]

208. Wang, H.; Zhang, C.; Chan, W.; Tiwari, S.; Rana, F. Ultrafast response of monolayer molybdenum disulfide photodetectors. *Nat. Commun.* **2015**, *6*, 8831. [CrossRef] [PubMed]

209. Lopez-Sanchez, O.; Lembke, D.; Kayci, M.; Radenovic, A.; Kis, A. Ultrasensitive photodetectors based on monolayer MoS$_2$. *Nat. Nanotechnol.* **2013**, *8*, 497–501. [CrossRef] [PubMed]

210. Zhang, W.; Huang, J.K.; Chen, C.H.; Chang, Y.H.; Cheng, Y.J.; Li, L.J. High-gain phototransistors based on a CVD MoS(2) monolayer. *Adv. Mater.* **2013**, *25*, 3456–3461. [CrossRef] [PubMed]

211. Perea-López, N.; Elías, A.L.; Berkdemir, A.; Castro-Beltran, A.; Gutiérrez, H.R.; Feng, S.; Lv, R.; Hayashi, T.; López-Urías, F.; Ghosh, S.; et al. Photosensor Device Based on Few-Layered WS$_2$Films. *Adv. Funct. Mater.* **2013**, *23*, 5511–5517. [CrossRef]

212. Zhang, W.; Chiu, M.H.; Chen, C.H.; Chen, W.; Li, L.J.; Wee, A.T. Role of metal contacts in high-performance phototransistors based on WSe$_2$ monolayers. *ACS Nano* **2014**, *8*, 8653–8661. [CrossRef] [PubMed]

213. Tsai, D.S.; Liu, K.K.; Lien, D.H.; Tsai, M.L.; Kang, C.F.; Lin, C.A.; Li, L.J.; He, J.H. Few-Layer MoS$_2$ with High Broadband Photogain and Fast Optical Switching for Use in Harsh Environments. *ACS Nano* **2013**, *7*, 3905–3911. [CrossRef] [PubMed]

214. Dung-Sheng, T.; Lien, D.H.; Tsai, M.L.; Su, S.H.; Chen, K.M.; Ke, J.J.; Yu, Y.C.; Li, L.J.; He, J.H. Trilayered MoS$_2$ Metal -Semiconductor-Metal Photodetectors: Photogain and Radiation Resistance. *IEEE J. Sel. Top. Quantum. Electron.* **2014**, *20*, 30–35. [CrossRef]

215. Yamaguchi, H.; Blancon, J.C.; Kappera, R.; Lei, S.; Najmaei, S.; Mangum, B.D.; Gupta, G.; Ajayan, P.M.; Lou, J.; Chhowalla, M.; et al. Spatially resolved photoexcited charge-carrier dynamics in phase-engineered monolayer MoS$_2$. *ACS Nano* **2015**, *9*, 840–849. [CrossRef] [PubMed]

216. Kang, D.-H.; Kim, M.S.; Shim, J.; Jeon, J.; Park, H.Y.; Jung, W.S.; Yu, H.Y.; Pang, C.H.; Lee, S.; Park, J.H. High-Performance Transition Metal Dichalcogenide Photodetectors Enhanced by Self-Assembled Monolayer Doping. *Adv. Funct. Mater.* **2015**, *25*, 4219–4227. [CrossRef]

217. Gan, X.T.; Shiue, R.J.; Gao, Y.D.; Meric, I.; Heinz, T.F.; Shepard, K.; Hone, J.; Assefa, S.; Englund, D. Chip-integrated ultrafast graphene photodetector with high responsivity. *Nat. Photonics* **2013**, *7*, 883–887. [CrossRef]

218. Mittendorff, M.; Winnerl, S.; Kamann, J.; Eroms, J.; Weiss, D.; Schneider, H.; Helm, M. Ultrafast graphene-based broadband THz detector. *Appl. Phys. Lett.* **2013**, *103*, 021113. [CrossRef]

219. Xu, X.; Gabor, N.M.; Alden, J.S.; van der Zande, A.M.; McEuen, P.L. Photo-thermoelectric effect at a graphene interface junction. *Nano Lett.* **2010**, *10*, 562–566. [CrossRef] [PubMed]

220. Zhang, Y.; Li, H.; Wang, L.; Wang, H.; Xie, X.; Zhang, S.L.; Liu, R.; Qiu, Z.J. Photothermoelectric and photovoltaic effects both present in MoS$_2$. *Sci. Rep.* **2015**, *5*, 7938. [CrossRef] [PubMed]

221. Buscema, M.; Barkelid, M.; Zwiller, V.; van der Zant, H.S.; Steele, G.A.; Castellanos-Gomez, A. Large and tunable photothermoelectric effect in single-layer MoS$_2$. *Nano Lett.* **2013**, *13*, 358–363. [CrossRef] [PubMed]

222. Konabe, S.; Yamamoto, T. Valley photothermoelectric effects in transition-metal dichalcogenides. *Phys. Rev. B* **2014**, *90*, 075430. [CrossRef]

223. Choi, W.; Cho, M.Y.; Konar, A.; Lee, J.H.; Cha, G.B.; Hong, S.C.; Kim, S.; Kim, J.; Jena, D.; Joo, J.; et al. High-detectivity multilayer MoS(2) phototransistors with spectral response from ultraviolet to infrared. *Adv. Mater.* **2012**, *24*, 5832–5836. [CrossRef] [PubMed]

224. Kim, H.C.; Kim, H.; Lee, J.U.; Lee, H.B.; Choi, D.H.; Lee, J.H.; Lee, W.H.; Jhang, S.H.; Park, B.H.; Cheong, H.; et al. Engineering Optical and Electronic Properties of WS$_2$ by Varying the Number of Layers. *ACS Nano* **2015**, *9*, 6854–6860. [CrossRef] [PubMed]

225. Jeong, H.; Oh, H.M.; Bang, S.; Jeong, H.J.; An, S.J.; Han, G.H.; Kim, H.; Yun, S.J.; Kim, K.K.; Park, J.C.; et al. Metal-Insulator-Semiconductor Diode Consisting of Two-Dimensional Nanomaterials. *Nano Lett.* **2016**, *16*, 1858–1862. [CrossRef] [PubMed]

226. Yu, S.H.; Lee, Y.; Jang, S.K.; Kang, J.; Jeon, J.; Lee, C.; Lee, J.Y.; Kim, H.; Hwang, E.; Lee, S.; et al. Dye-sensitized MoS$_2$ photodetector with enhanced spectral photoresponse. *ACS Nano* **2014**, *8*, 8285–8291. [CrossRef] [PubMed]

227. Kufer, D.; Nikitskiy, I.; Lasanta, T.; Navickaite, G.; Koppens, F.H.; Konstantatos, G. Hybrid 2D-0D MoS$_2$ -PbS quantum dot photodetectors. *Adv. Mater.* **2015**, *27*, 176–180. [CrossRef] [PubMed]

228. Kang, D.H.; Pae, S.R.; Shim, J.; Yoo, G.; Jeon, J.; Leem, J.W.; Yu, J.S.; Lee, S.; Shin, B.; Park, J.H. An Ultrahigh-Performance Photodetector based on a Perovskite-Transition-Metal-Dichalcogenide Hybrid Structure. *Adv. Mater.* **2016**, *28*, 7799–7806. [CrossRef] [PubMed]

229. Splendiani, A.; Sun, L.; Zhang, Y.; Li, T.; Kim, J.; Chim, C.Y.; Galli, G.; Wang, F. Emerging photoluminescence in monolayer MoS_2. *Nano Lett.* **2010**, *10*, 1271–1275. [CrossRef] [PubMed]

230. Eda, G.; Yamaguchi, H.; Voiry, D.; Fujita, T.; Chen, M.; Chhowalla, M. Photoluminescence from chemically exfoliated MoS_2. *Nano Lett.* **2011**, *11*, 5111–5116. [CrossRef] [PubMed]

231. Gutierrez, H.R.; Perea-Lopez, N.; Elias, A.L.; Berkdemir, A.; Wang, B.; Lv, R.; Lopez-Urias, F.; Crespi, V.H.; Terrones, H.; Terrones, M. Extraordinary room-temperature photoluminescence in triangular WS_2 monolayers. *Nano Lett.* **2013**, *13*, 3447–3454. [CrossRef] [PubMed]

232. Tonndorf, P.; Schmidt, R.; Bottger, P.; Zhang, X.; Borner, J.; Liebig, A.; Albrecht, M.; Kloc, C.; Gordan, O.; Zahn, D.R.T.; et al. Photoluminescence emission and Raman response of monolayer MoS_2, $MoSe_2$, and WSe_2. *Opt. Express* **2013**, *21*, 4908–4916. [CrossRef] [PubMed]

233. Mak, K.F.; Lee, C.; Hone, J.; Shan, J.; Heinz, T.F. Atomically Thin MoS_2: A New Direct-Gap Semiconductor. *Phys. Rev. Lett.* **2010**, *105*, 136805. [CrossRef] [PubMed]

234. Salehzadeh, O.; Tran, N.H.; Liu, X.; Shih, I.; Mi, Z. Exciton kinetics, quantum efficiency, and efficiency droop of monolayer MoS(2) light-emitting devices. *Nano Lett.* **2014**, *14*, 4125–4130. [CrossRef] [PubMed]

235. Salehzadeh, O.; Djavid, M.; Tran, N.H.; Shih, I.; Mi, Z. Optically Pumped Two-Dimensional MoS_2 Lasers Operating at Room-Temperature. *Nano Lett.* **2015**, *15*, 5302–5306. [CrossRef] [PubMed]

236. Gan, X.; Gao, Y.; Fai Mak, K.; Yao, X.; Shiue, R.J.; van der Zande, A.; Trusheim, M.E.; Hatami, F.; Heinz, T.F.; Hone, J.; et al. Controlling the spontaneous emission rate of monolayer MoS_2 in a photonic crystal nanocavity. *Appl. Phys. Lett.* **2013**, *103*, 181119. [CrossRef] [PubMed]

237. Wu, S.F.; Buckley, S.; Jones, A.M.; Ross, J.S.; Ghimire, N.J.; Yan, J.Q.; Mandrus, D.G.; Yao, W.; Hatami, F.; Vuckovic, J.; et al. Control of two-dimensional excitonic light emission via photonic crystal. *2D Mater.* **2014**, *1*, 011001. [CrossRef]

238. Schwarz, S.; Dufferwiel, S.; Walker, P.M.; Withers, F.; Trichet, A.A.; Sich, M.; Li, F.; Chekhovich, E.A.; Borisenko, D.N.; Kolesnikov, N.N.; et al. Two-dimensional metal-chalcogenide films in tunable optical microcavities. *Nano Lett.* **2014**, *14*, 7003–7008. [CrossRef] [PubMed]

239. Liu, X.; Galfsky, T.; Sun, Z.; Xia, F.; Lin, E.-C.; Lee, Y.-H.; Kéna-Cohen, S.; Menon, V.M. Strong light–matter coupling in two-dimensional atomic crystals. *Nat. Photonics* **2014**, *9*, 30–34. [CrossRef]

240. Wu, S.; Buckley, S.; Schaibley, J.R.; Feng, L.; Yan, J.; Mandrus, D.G.; Hatami, F.; Yao, W.; Vuckovic, J.; Majumdar, A.; et al. Monolayer semiconductor nanocavity lasers with ultralow thresholds. *Nature* **2015**, *520*, 69–72. [CrossRef] [PubMed]

241. Mak, K.F.; He, K.; Lee, C.; Lee, G.H.; Hone, J.; Heinz, T.F.; Shan, J. Tightly bound trions in monolayer MoS_2. *Nat. Mater.* **2013**, *12*, 207–211. [CrossRef] [PubMed]

242. Ye, Z.; Cao, T.; O'Brien, K.; Zhu, H.; Yin, X.; Wang, Y.; Louie, S.G.; Zhang, X. Probing excitonic dark states in single-layer tungsten disulphide. *Nature* **2014**, *513*, 214–218. [CrossRef] [PubMed]

243. He, K.; Kumar, N.; Zhao, L.; Wang, Z.; Mak, K.F.; Zhao, H.; Shan, J. Tightly bound excitons in monolayer WSe(2). *Phys. Rev. Lett.* **2014**, *113*, 026803. [CrossRef] [PubMed]

244. Ye, Y.; Wong, Z.J.; Lu, X.F.; Ni, X.J.; Zhu, H.Y.; Chen, X.H.; Wang, Y.; Zhang, X. Monolayer excitonic laser. *Nat. Photonics* **2015**, *9*, 733–737. [CrossRef]

245. Song, H.S.; Li, S.L.; Gao, L.; Xu, Y.; Ueno, K.; Tang, J.; Cheng, Y.B.; Tsukagoshi, K. High-performance top-gated monolayer SnS2 field-effect transistors and their integrated logic circuits. *Nanoscale* **2013**, *5*, 9666–9670. [CrossRef] [PubMed]

246. Tosun, M.; Chuang, S.; Fang, H.; Sachid, A.B.; Hettick, M.; Lin, Y.; Zeng, Y.; Javey, A. High-gain inverters based on WSe_2 complementary field-effect transistors. *ACS Nano* **2014**, *8*, 4948–4953. [CrossRef] [PubMed]

247. Cho, A.J.; Park, K.C.; Kwon, J.Y. A high-performance complementary inverter based on transition metal dichalcogenide field-effect transistors. *Nanoscale Res. Lett.* **2015**, *10*, 115. [CrossRef] [PubMed]

248. Jeon, P.J.; Kim, J.S.; Lim, J.Y.; Cho, Y.; Pezeshki, A.; Lee, H.S.; Yu, S.; Min, S.W.; Im, S. Low Power Consumption Complementary Inverters with n-MoS_2 and p-WSe_2 Dichalcogenide Nanosheets on Glass for Logic and Light-Emitting Diode Circuits. *ACS Appl. Mater. Interfaces* **2015**, *7*, 22333–22340. [CrossRef] [PubMed]

249. Wang, H.; Yu, L.; Lee, Y.H.; Shi, Y.; Hsu, A.; Chin, M.L.; Li, L.J.; Dubey, M.; Kong, J.; Palacios, T. Integrated circuits based on bilayer MoS(2) transistors. *Nano Lett.* **2012**, *12*, 4674–4680. [CrossRef] [PubMed]

250. Lipatov, A.; Sharma, P.; Gruverman, A.; Sinitskii, A. Optoelectrical Molybdenum Disulfide (MoS_2)-Ferroelectric Memories. *ACS Nano* **2015**, *9*, 8089–8098. [CrossRef] [PubMed]

251. Zhang, E.; Wang, W.; Zhang, C.; Jin, Y.; Zhu, G.; Sun, Q.; Zhang, D.W.; Zhou, P.; Xiu, F. Tunable charge-trap memory based on few-layer MoS_2. *ACS Nano* **2015**, *9*, 612–619. [CrossRef] [PubMed]

252. Bertolazzi, S.; Krasnozhon, D.; Kis, A. Nonvolatile memory cells based on MoS_2/graphene heterostructures. *ACS Nano* **2013**, *7*, 3246–3252. [CrossRef] [PubMed]

253. Choi, M.S.; Lee, G.H.; Yu, Y.J.; Lee, D.Y.; Lee, S.H.; Kim, P.; Hone, J.; Yoo, W.J. Controlled charge trapping by molybdenum disulphide and graphene in ultrathin heterostructured memory devices. *Nat. Commun.* **2013**, *4*, 1624. [CrossRef] [PubMed]

254. Vu, Q.A.; Shin, Y.S.; Kim, Y.R.; Nguyen, V.L.; Kang, W.T.; Kim, H.; Luong, D.H.; Lee, I.M.; Lee, K.; Ko, D.S.; et al. Two-terminal floating-gate memory with van der Waals heterostructures for ultrahigh on/off ratio. *Nat. Commun.* **2016**, *7*, 12725. [CrossRef] [PubMed]

255. Roy, K.; Padmanabhan, M.; Goswami, S.; Sai, T.P.; Ramalingam, G.; Raghavan, S.; Ghosh, A. Graphene-MoS_2 hybrid structures for multifunctional photoresponsive memory devices. *Nat. Nanotechnol.* **2013**, *8*, 826–830. [CrossRef] [PubMed]

256. Lee, Y.T.; Lee, J.; Ju, H.; Lim, J.A.; Yi, Y.; Choi, W.K.; Hwang, D.K.; Im, S. Nonvolatile Charge Injection Memory Based on Black Phosphorous 2D Nanosheets for Charge Trapping and Active Channel Layers. *Adv. Funct. Mater.* **2016**, *26*, 5701–5707. [CrossRef]

257. Liu, H.L.; Shen, C.C.; Su, S.H.; Hsu, C.L.; Li, M.Y.; Li, L.J. Optical properties of monolayer transition metal dichalcogenides probed by spectroscopic ellipsometry. *Appl. Phys. Lett.* **2014**, *105*, 201905. [CrossRef]

electronics

MDPI

Article

High Throughput Characterization of Epitaxially Grown Single-Layer MoS$_2$

Foad Ghasemi [1,2], Riccardo Frisenda [1,*], Dumitru Dumcenco [3,4], Andras Kis [3,4], David Perez de Lara [1] and Andres Castellanos-Gomez [1,5,*]

[1] Instituto Madrileño de Estudios Avanzados en Nanociencia (IMDEA-nanociencia), Campus de Cantoblanco, E-28049 Madrid, Spain; foad.ghasemi@imdea.org (F.G.); david.perezdelara@imdea.org (D.P.d.L.)
[2] Nanoelectronic Lab, School of Electrical and Computer Engineering, University of Tehran, 14399–56191 Tehran, Iran
[3] Electrical Engineering Institute, École Polytechnique Fédérale de Lausanne (EPFL), CH-1015 Lausanne, Switzerland; dumitru.dumcenco@epfl.ch (D.D.); andras.kis@epfl.ch (A.K.)
[4] Institute of Materials Science and Engineering, École Polytechnique Fédérale de Lausanne (EPFL), CH-1015 Lausanne, Switzerland
[5] Instituto de Ciencia de los Materiales de Madrid (ICMM-CSIC), E-28049 Madrid, Spain
* Correspondence: riccardo.frisenda@imdea.org (R.F.); andres.castellanos@csic.es (A.C.-G.); Tel.: +34-91-299-8770 (R.F. & A.C.-G.)

Academic Editors: Yoke Khin Yap and Zhixian Zhou
Received: 21 February 2017; Accepted: 28 March 2017; Published: 31 March 2017

Abstract: The growth of single-layer MoS$_2$ with chemical vapor deposition is an established method that can produce large-area and high quality samples. In this article, we investigate the geometrical and optical properties of hundreds of individual single-layer MoS$_2$ crystallites grown on a highly-polished sapphire substrate. Most of the crystallites are oriented along the terraces of the sapphire substrate and have an area comprised between 10 μm^2 and 60 μm^2. Differential reflectance measurements performed on these crystallites show that the area of the MoS$_2$ crystallites has an influence on the position and broadening of the B exciton while the orientation does not influence the A and B excitons of MoS$_2$. These measurements demonstrate that differential reflectance measurements have the potential to be used to characterize the homogeneity of large-area chemical vapor deposition (CVD)-grown samples.

Keywords: two dimensional materials; chemical vapor deposition; MoS$_2$; reflectance; exciton

1. Introduction

Rapidly after the first works on mechanically exfoliated MoS$_2$, part of the efforts of the scientific community working on two-dimensional (2D) materials focused on developing synthesis methods that could provide large-area single-layer MoS$_2$ [1–6]. Chemical vapor deposition (CVD)-based growth methods, which were already successfully used in the growth of graphene, are nowadays standard techniques to grow large-area samples of single-layer MoS$_2$ and other transition metal dichalcogenides [7–12]. In fact, CVD samples have shown remarkable electronic and optical properties approaching those of mechanically exfoliated material [13–16].

Although some techniques such as Raman spectroscopy and photoluminescence (PL) mapping have been used to investigate the structural and electronic properties of 2D layers based CVD growth [17–21], a thorough statistical study of the uniformity of the as-grown samples is still somewhat lacking. In this work we use micro-reflectance spectroscopy measurements to investigate the differential reflectance spectra of hundreds of CVD-grown MoS$_2$ flakes grown on a highly-polished sapphire substrate. This technique is a very fast and non-destructive characterization tool that allows us to measure a large number of spectra in different sample locations and in a small amount of time. This can provide useful statistical information about the homogeneity of the optical properties of the samples.

2. Materials and Methods

The CVD growth of MoS_2 is based on the gas-phase reaction between MoO_3 ($\geq$99.998% Alfa Aesar) and high-pure sulfur evaporated from the solid phase ($\geq$99.99% purity, Sigma Aldrich), for detailed information about the growth see reference [22]. Figure 1a shows an optical microscopy image (epi-illumination mode) of a 60×60 μm region of the sample containing single-layer MoS_2 crystallites grown onto the sapphire surface. Most of the crystallites have approximately an equilateral triangular shape and they appear to be oriented following preferential orientation with respect to the sapphire lattice as previously reported [22,23]. As can be seen in the right part of the picture, some of the crystallites can coalesce and merge. In this article we focus only on the isolated triangular MoS_2 crystallites for our analysis. To extract statistical information about the geometry of the MoS_2 crystallites, we collected 50 optical images (108×86 μm) of the sample surface, taken in different regions. From each image we extract the area and the orientation of the single-layer MoS_2 triangular crystallites.

The optical characterization of the single-layer MoS_2 crystallites has been carried out with differential reflectance spectroscopy. More details on the technique and about the experimental setup can be found in Reference [24] but briefly we illuminate a 60 μm spot on the sample surface with white light and we collect the light reflected from an area of the sample of 2 μm in diameter. Subtracting the intensity reflected by the bare substrate from the intensity of the light reflected by the MoS_2 flake, and normalizing the result by the intensity reflected by the MoS_2, one can extract a differential reflectance spectrum, which in the case of a thin film is proportional to the absorption of the thin film [25,26].

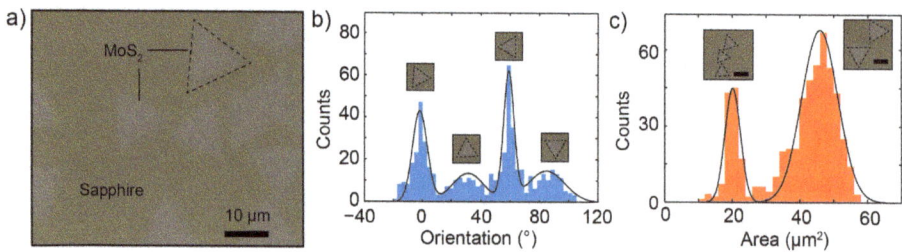

Figure 1. (**a**) Optical image of the grown single layer MoS_2 with triangle shape (the dash-line highlights single crystallites) on sapphire; (**b**) Orientation histogram of 550 MoS_2 crystallites. The black line is a fit to four Gaussian peaks. Inset: orientation of four crystallites corresponding respectively to an angle of 0°, 30°, 60° and 90°; (**c**) Area histogram of 550 MoS_2 flakes. The black line is a fit to two Gaussian peaks. Inset: optical pictures of two different regions of the sample showing respectively small-area crystallites and large-area ones. The scale bar is in both cases 10 μm.

3. Results and Discussions

We analyzed 550 MoS_2 triangular flakes and we built histograms of the area and the orientation. Figure 1b shows a histogram of the triangle orientation with the orientation angle defined as in the pictures in the inset. The histogram presents two main peaks located at 0° and 60° and two secondary peaks centered at 30° and 90°. We fit the histogram to four Gaussian peaks and we find that 380 flakes (69% of the total population) are oriented around 0° or around 60° while 170 (31%) flakes have an angle around 30° or 90°. The presence of clear peaks in the orientation histogram indicates the presence of preferential orientation populations of MoS_2 triangles on the surface of the sample. Since the single-crystal sapphire substrate has a two-fold symmetry (it is invariant for a rotation of 180°), the populations with orientation of 0° and 60° are physically equivalent and are the most probable ones (as well as two other orientations at 30° and 90°). The presence of these two different populations can be explained by the growth dynamic of CVD MoS_2 onto sapphire [23,27] with the most probable configuration, the 0°/60°, being the one where MoS_2 islands and the sapphire lattice are commensurate.

Similarly to the previous analysis, we extracted the area of 550 MoS_2 islands and we built a histogram of the triangles area, which is shown in Figure 1c. The histogram exhibits two peaks, one centered at 20 μm^2 and the other at 47 μm^2. We fitted two Gaussian peaks to the histogram and from the peaks areas we find that 118 flakes (22% of the total population) have a small area (around 20 μm^2) while 431 flakes (78%) have a large area (around 47 μm^2). We attribute this difference in the triangle size distribution to be originated by spatial variations of the temperature and/or precursors flow during the growth process over the sapphire.

We further characterize the homogeneity of the sample by means of micro-reflectance spectroscopy. Figure 2a shows a differential reflectance spectrum of an MoS_2 island. The spectrum increases for increasing energy and is characterized by two prominent peaks centered at 1.877 eV and 2.016 eV which correspond to excitonic resonances originating from direct valence to conduction band transitions at the K point [26,28,29], as schematically depicted in the inset of Figure 2. The presence of two features is due to the spin-orbit splitting mostly of the valance band and thus the difference in energy between these peaks gives a good estimation of the spin-orbit interaction in MoS_2.

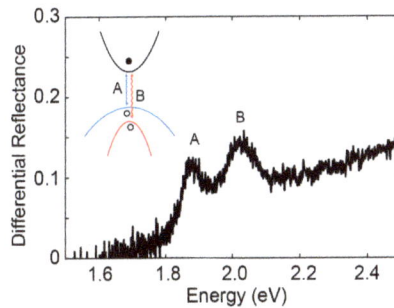

Figure 2. Differential reflectance spectrum of a single-layer MoS_2 triangle showing A and B exciton peaks. Inset: schematic drawing of the band diagram of single-layer MoS_2 indicating the A and B excitons at the K point.

For each one of the 550 MoS_2 crystallites previously identified, we carried out differential reflectance measurements. From each spectrum, we extracted the energy and the full-width-at-half-maximum (FWHM) values of the A and B excitons by fitting Gaussian curves on the obtained spectra. Figure 3 shows the histograms built up with those values. The energy of the A and B exciton and their broadening have variations across the CVD sample as evidenced by the finite width of all four histograms. The shape of the histograms, which is not that of a simple Gaussian curve, suggests that random fluctuations alone cannot explain the origin of the exciton variability. Possibly additional mechanisms, like the interaction with the substrate, may play a role in the spectrum-to-spectrum variability. We extract the maximum of each histogram and we find that the average A and B excitons energy is respectively 1.876 eV and 2.016 eV, while the FWHM of the A and B excitons is in average 71 meV and 124 meV respectively. Comparing these values with values that we found previously for mechanical exfoliated single-layer MoS_2 [24] (A 1.90 eV, B 2.04 eV) we see that an overall blueshift of 24 meV is present in both the A and B exciton position of CVD-grown MoS_2. The difference between the A and B mean energy is 140 meV, which is comparable with previous studies of single-layer MoS_2 [26,28].

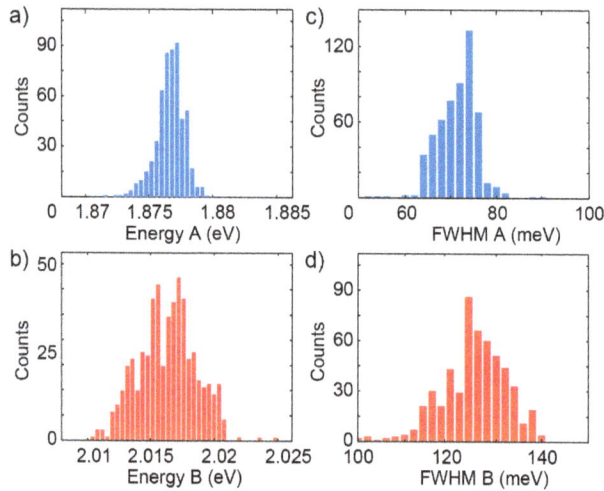

Figure 3. Extracted data from Differential reflectance measurement: (**a,b**) Histograms of the A and B exciton energy for 550 flakes. (**c,d**) Histograms of the full width at half maximum of the A and B excitons peaks.

In order to get a deeper insight in this spectrum-to-spectrum fluctuations observed in CVD MoS_2 sample, we investigate the correlations between the A and B exciton energy and FWHM and the geometrical parameters of the MoS_2 islands such as the area and the orientation. Figure 4a shows the correlation between the A exciton energy with the triangle orientation. The bottom plot represents a scatter plot of the A exciton energy as a function of the island orientation (black dots) on top of a color map which depends on the data points density [30]. The scatter plot shows two clouds of data points corresponding to the most probable orientation (0° or equivalently 60°) and the least probable orientation (30° or 90°). Both clouds appear to be centered at the same A energy suggesting that the orientation does not affect the excitons in CVD MoS_2. In the top panel we have included two one-dimensional conditional histograms of the A exciton energy, each constructed only from the islands belonging to one of the two orientation populations. The conditional histograms of the two populations appear similar and by fitting each histogram to a Gaussian peak we find that the difference in energy of the A exciton between the 0° and 30° orientation is (0.1 ± 0.1) meV, which can be considered zero within the experimental uncertainty. The uncertainty is calculated by propagating the standard errors of the peaks center. Similar conclusions hold for the energy of exciton B, presented in Figure 4b, the difference in the energy of the B exciton between triangles with 0° and 30° orientation is (0.4 ± 0.3) meV. We performed the same correlation analysis for the broadening (FWHM) of the A and B peaks. Figure 4c,d report similar plots for the excitons peak FWHM. Again we do not observe a correlation between the broadening of the excitons and the orientation of the MoS_2 islands and finding a negligible difference between the two distributions, (0.4 ± 0.5) meV in the case of exciton A and (1.5 ± 0.9) meV in the case of B. This small difference in average excitons energy for the different MoS_2 orientations might be due to the different MoS_2/sapphire interaction that could induce a different strain in the MoS_2 layers with the 0° and 30° orientations [11,31].

Finally, we perform the same correlation analysis discussed in Figure 4 to the MoS_2 islands area instead of the orientation. Figure 5a,b show the correlation between the A and B exciton energy with the triangle area. The scatter plot shows two clouds of data points corresponding to the smallest triangle area ($\approx$20 μm^2) and to the largest area ($\approx$47 μm^2). Interestingly, while for exciton A both clouds appear to be centered at the same energy, in the case of exciton B the points related to the smaller area are shifted toward smaller energy. In the top panel we have included one-dimensional conditional histograms of the A (B) exciton energy, which are constructed only from the islands belonging to one of the two area populations. The conditional histograms of the two populations appear similar for the A exciton energy, and by fitting each histogram to a Gaussian peak we find that the difference in energy of the A exciton, between the small and the large area populations, is (0.4 ± 0.1) meV. In the case of the B exciton instead we find that the smaller area islands have an energy (3.0 ± 0.2) meV smaller than the large area ones. We performed the same correlation analysis for the broadening (FWHM) of the A and B peaks and Figure 5c,d report similar plots for the excitons peak FWHM. We do not observe a correlation between the broadening of the A exciton and the area of the MoS_2 islands finding a negligible difference of (0.5 ± 0.5) meV between the two distributions. The B exciton FWHM shows instead a marked difference of (7.5 ± 1.0) meV between large and small areas.

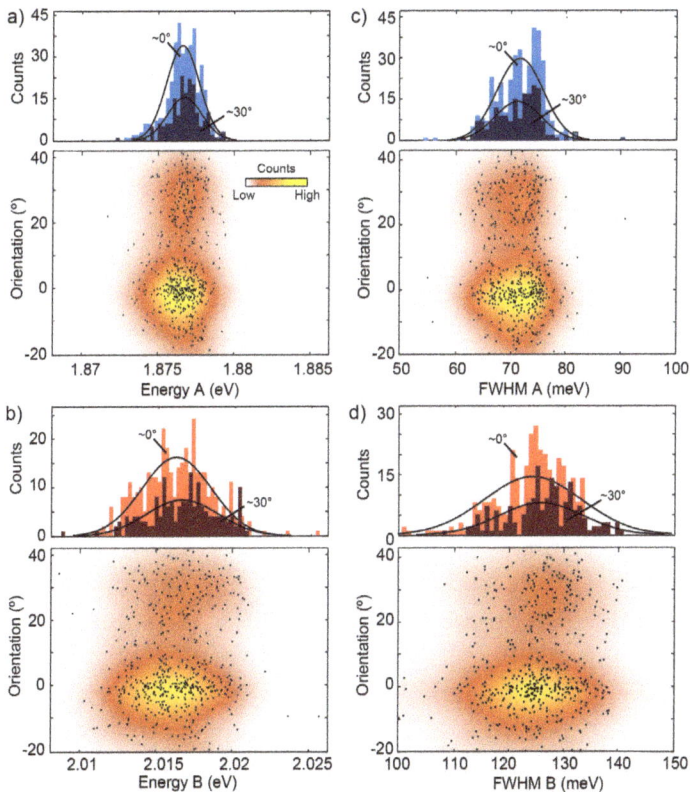

Figure 4. (**a**) Density plot of the triangle orientation as a function of the A exciton energy. (top) Conditional histograms of the A exciton energy for triangles with orientation $0° \pm 10°$ (light blue) and triangles with orientation $30° \pm 10°$ (dark blue); (**b**) Same as panel (**a**) for exciton B; (**c**) Density plot of the triangle orientation as a function of the A exciton peak FWHM. (top) Conditional histograms of the A exciton peak FWHM for triangles with orientation $0° \pm 10°$ (light blue) and triangles with orientation $30° \pm 10°$ (dark blue); (**d**) Same as panel (**c**) for exciton B.

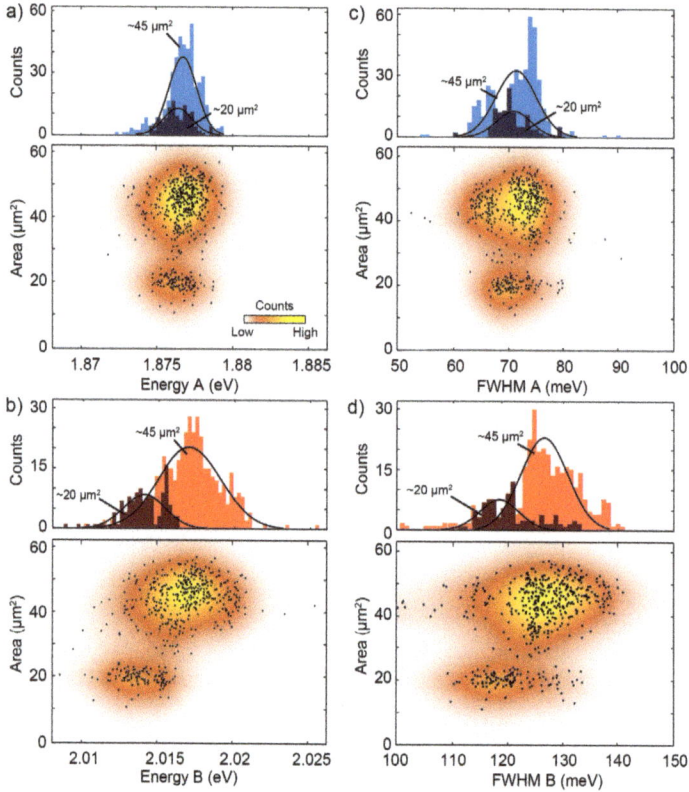

Figure 5. (**a**) Density plot of the triangle orientation as a function of the A exciton energy. (top) Conditional histograms of the A exciton energy for triangles with area between 15 μm^2 and 25 μm^2 (light blue) and triangles with area between 35 μm^2 and 55 μm^2 (dark blue); (**b**) Same as panel (**a**) for exciton B; (**c**) Density plot of the triangle orientation as a function of the A exciton peak FWHM. (top) Conditional histograms of the A exciton peak FWHM for triangles with area between 15 μm^2 and 25 μm^2 (light blue) and triangles with area between 35 μm^2 and 55 μm^2 (dark blue); (**d**) Same as panel (**c**) for exciton B.

4. Discussions

The results of the previous section indicate that the orientation of the MoS$_2$ crystallites in respect to the sapphire substrate does not affect the A and B excitons energy or broadening. Conversely, the area of the crystal shows a correlation with the position and broadening of the B exciton. From literature it is known that CVD-grown MoS$_2$ can show non-homogeneous strain [21,32,33]. One of the possible scenarios is that in our sample the CVD growth method introduces strain in the MoS$_2$ lattice, thus modifying the bands of MoS$_2$ and the excitons. The area of the MoS$_2$ crystallite can determine the amount of strain transfer with the substrate and thus influence the band structure. A full understanding of this effect is still missing, but it deserves further investigation in future works.

To conclude, we employed micro-reflectance spectroscopy to statistically study the variation in the optical properties of single-layer MoS$_2$ triangular crystallites grown by CVD on a sapphire substrate. We measured the spectra at 550 positions in the sample and we found the distributions of the A and B exciton energies and broadening. Interestingly, we found a correlation between the B exciton energy and broadening and the MoS$_2$ crystallite area. We thus demonstrate that this technique has the potential to be used to characterize the homogeneity of large area CVD grown samples.

Acknowledgments: Andres Castellanos-Gomez acknowledges support from the European Commission (Graphene Flagship: contract CNECTICT-604391), the MINECO (Ramón y Cajal 2014 program RYC-2014-01406 and program MAT2014-58399-JIN) and the Comunidad de Madrid (MAD2D-CM program S2013/MIT-3007). Riccardo Frisenda acknowledges support from The Netherlands Organisation for Scientific Research (NWO, Rubicon 680-50-1515). David Perez de Lara acknowledges support from the MINECO (program FIS2015-67367-C2-1-p). Andras Kis and Dimitri Dumcenco acknowledge funding from Swiss SNF Sinergia Grant No. 147607.

Author Contributions: F.G. collected the differential reflectance spectra and analyzed the optical images. F.G., R.F. and A.C.G. analyzed the experimental data. A.K. and D.D. performed the MoS$_2$ CVD growth. A.C.G. and D.P.d.L. designed the experiment. All the authors contributed to discussions and writing of the manuscript.

Conflicts of Interest: The authors declare no conflict of interest.

References

1. Liu, K.K.; Zhang, W.; Lee, Y.H.; Lin, Y.C.; Chang, M.T.; Su, C.Y.; Chang, C.S.; Li, H.; Shi, Y.; Zhang, H.; et al. Growth of Large-Area and Highly Crystalline MoS$_2$ Thin Layers on Insulating Substrates. *Nano Lett.* **2012**, *12*, 1538–1544. [CrossRef] [PubMed]

2. Zhan, Y.; Liu, Z.; Najmaei, S.; Ajayan, P.M.; Lou, J. Large-Area Vapor-Phase Growth and Characterization of MoS$_2$ Atomic Layers on a SiO$_2$ Substrate. *Small* **2012**, *8*, 966–971. [CrossRef] [PubMed]

3. Yu, Y.; Li, C.; Liu, Y.; Su, L.; Zhang, Y.; Cao, L. Controlled Scalable Synthesis of Uniform, High-Quality Monolayer and Few-layer MoS$_2$ Films. *Sci. Rep.* **2013**, *3*, 1866. [CrossRef] [PubMed]

4. Laskar, M.R.; Ma, L.; Kannappan, S.; Park, P.S.; Krishnamoorthy, S.; Nath, D.N.; Lu, W.; Wu, Y.; Rajan, S. Large area single crystal (0001) oriented MoS$_2$. *Appl. Phys. Lett.* **2013**, *102*, 252108. [CrossRef]

5. Butler, S.Z.; Hollen, S.M.; Cao, L.; Cui, Y.; Gupta, J.A.; Gutiérrez, H.R.; Heinz, T.F.; Hong, S.S.; Huang, J.; Ismach, A.F.; et al. Progress, challenges, and opportunities in two-dimensional materials beyond graphene. *ACS Nano* **2013**, *7*, 2898–2926. [CrossRef] [PubMed]

6. Jariwala, D.; Sangwan, V.K.; Lauhon, L.J.; Marks, T.J.; Hersam, M.C. Emerging device applications for semiconducting two-dimensional transition metal dichalcogenides. *ACS Nano* **2014**, *8*, 1102–1120. [CrossRef] [PubMed]

7. Zhang, Y.; Zhang, Y.; Ji, Q.; Ju, J.; Yuan, H.; Shi, J.; Gao, T.; Ma, D.; Liu, M.; Chen, Y.; et al. Controlled Growth of High-Quality Monolayer WS$_2$ Layers on Sapphire and Imaging Its Grain Boundary. *ACS Nano* **2013**, *7*, 8963–8971. [CrossRef] [PubMed]

8. Wu, S.; Huang, C.; Aivazian, G.; Ross, J.S.; Cobden, D.H.; Xu, X. Vapor–Solid Growth of High Optical Quality MoS$_2$ Monolayers with Near-Unity Valley Polarization. *ACS Nano* **2013**, *7*, 2768–2772. [CrossRef] [PubMed]

9. Wang, X.; Feng, H.; Wu, Y.; Jiao, L. Controlled Synthesis of Highly Crystalline MoS$_2$ Flakes by Chemical Vapor Deposition. *J. Am. Chem. Soc.* **2013**, *135*, 5304–5307. [CrossRef] [PubMed]

10. Kobayashi, Y.; Sasakim, S.; Mori, S.; Hibino, H.; Liu, Z.; Watanabe, K.; Taniguchi, T.; Suenaga, K.; Maniwa, Y.; Miyata, Y. Growth and Optical Properties of High-Quality Monolayer WS$_2$ on Graphite. *ACS Nano* **2015**, *9*, 4056–4063. [CrossRef] [PubMed]

11. Liu, H.F.; Wong, S.L.; Chi, D.Z. CVD Growth of MoS$_2$-based Two-dimensional Materials. *Chem. Vap. Depos.* **2015**, *21*, 241–259. [CrossRef]

12. Lee, Y.H.; Zhang, X.Q.; Zhang, W.; Chang, M.T.; Lin, C.T.; Chang, K.D.; Yu, Y.C.; Wang, J.T.W.; Chang, C.S.; Li, L.J.; et al. Synthesis of Large-Area MoS$_2$ Atomic Layers with Chemical Vapor Deposition. *Adv. Mater.* **2012**, *24*, 2320–2325. [CrossRef] [PubMed]

13. Schmidt, H.; Wang, S.; Chu, L.; Toh, M.; Kumar, R.; Zhao, W.; Castro Neto, A.H.; Martin, J.; Adam, S.; Ozyilmaz, B.; et al. Transport Properties of Monolayer MoS$_2$ Grown by Chemical Vapor Deposition. *Nano Lett.* **2014**, *14*, 1909–1913. [CrossRef] [PubMed]

14. Yi, Z.; Yang, J.; Zhang, S.; Mokhtar, S.; Pei, J.; Wang, X.; Lu, Y. Strongly enhanced photoluminescence in nanostructured monolayer MoS2 by chemical vapor deposition. *Nanotechnology* **2016**, *27*, 135706.

15. Bilgin, I.; Liu, F.; Vargas, A.; Winchester, A.; Man, M.K.L.; Upmanyu, M.; Dani, K.M.; Gupta, G.; Talapatra, S.; Mohite, A.D.; et al. Chemical Vapor Deposition Synthesized Atomically Thin Molybdenum Disulfide with Optoelectronic-Grade Crystalline Quality. *ACS Nano* **2015**, *9*, 8822–8832. [CrossRef] [PubMed]

16. Van der Zande, A.M.; Huang, P.Y.; Chenet, D.A.; Berkelbach, T.C.; You, Y.; Lee, G.H.; Heinz, T.F.; Reichman, D.R.; Muller, D.A.; Hone, J.C. Grains and grain boundaries in highly crystalline monolayer molybdenum disulphide. *Nat. Mater.* **2013**, *12*, 554–561. [CrossRef] [PubMed]

17. McCreary, K.M.; Hanbicki, A.T.; Robinson, J.T.; Cobas, E.; Culbertson, J.C.; Friedman, A.L.; Jernigan, G.G.; Jonker, B.T. Large-Area Synthesis of Continuous and Uniform MoS_2 Monolayer Films on Graphene. *Adv. Funct. Mater.* **2014**, *24*, 6449–6454. [CrossRef]
18. Senthilkumar, V.; Tam, L.C.; Kim, Y.S.; Sim, Y.; Seong, M.J.; Jang, J.I. Direct vapor phase growth process and robust photoluminescence properties of large area MoS_2 layers. *Nano Res.* **2014**, *7*, 1759–1768. [CrossRef]
19. Plechinger, G.; Mann, J.; Preciado, E.; Barroso, D.; Nguyen, A.; Eroms, J.; Schuller, C.; Bartels, L.; Korn, T. A direct comparison of CVD-grown and exfoliated MoS_2 using optical spectroscopy. *Semicond. Sci. Technol.* **2014**, *29*, 064008. [CrossRef]
20. Kim, I.S.; Sangwan, V.K.; Jariwala, D.; Wood, J.D.; Park, S.; Chen, K.S.; Shi, F.; Ruiz-Zepeda, F.; Ponce, A.; Jose-Yacaman, M.; et al. Influence of stoichiometry on the optical and electrical properties of chemical vapor deposition derived MoS_2. *ACS Nano* **2014**, *8*, 10551–10558. [CrossRef] [PubMed]
21. Liu, Z.; Amani, M.; Najmaei, S.; Xu, Q.; Zou, X.; Zhou, W.; Yu, T.; Qiu, C.; Glen Birdwell, A.; Crowne, F.J.; et al. Strain and structure heterogeneity in MoS_2 atomic layers grown by chemical vapour deposition. *Nat. Commun.* **2014**, *5*, 5246. [CrossRef] [PubMed]
22. Dumcenco, D.; Ovchinnikov, D.; Marinov, K.; Lazic, P.; Gibertini, M.; Marzari, N.; Lopez Sanchez, O.; Kung, Y.C.; Krasnozhon, D.; Chen, M.W.; et al. Large-Area Epitaxial Monolayer MoS_2. *ACS Nano* **2015**, *9*, 4611–4620. [CrossRef] [PubMed]
23. Ji, Q.; Kan, M.; Zhang, Y.; Guo, Y.; Ma, D.; Shi, J.; Sun, Q.; Chen, Q.; Zhang, Y.; Liu, Z. Unravelling Orientation Distribution and Merging Behavior of Monolayer MoS_2 Domains on Sapphire. *Nano Lett.* **2015**, *15*, 198–205. [CrossRef] [PubMed]
24. Frisenda, R.; Niu, Y.; Gant, P.; Molina-Mendoza, A.J.; Schmidt, R.; Bratschitsch, R.; Liu, J.; Fu, L.; Dumcenco, D.; Kis, A.; et al. Micro-reflectance and transmittance spectroscopy: A versatile and powerful tool to characterize 2D materials. *J. Phys. D Appl. Phys.* **2017**, *50*, 074002. [CrossRef]
25. Dhakal, K.P.; Duong, D.L.; Lee, J.; Nam, H.; Kim, M.; Kan, M.; Lee, Y.H.; Kim, J. Confocal absorption spectral imaging of MoS_2: Optical transitions depending on the atomic thickness of intrinsic and chemically doped MoS_2. *Nanoscale* **2014**, *6*, 13028–13035. [CrossRef] [PubMed]
26. Kozawa, D.; Kumar, R.; Carvalho, A.; Amara, K.K.; Zhao, W.; Wang, S.; Toh, M.; Ribeiro, R.M.; Castro Neto, A.H.; Matsuda, K.; et al. Photocarrier relaxation pathway in two-dimensional semiconducting transition metal dichalcogenides. *Nat. Commun.* **2014**, *5*, 4543. [CrossRef] [PubMed]
27. Liang, T.; Xie, S.; Huang, Z.; Fu, W.; Cai, Y.; Yang, X.; Chen, H.; Ma, X.; Iwai, H.; Fujita, D.; et al. Elucidation of Zero-Dimensional to Two-Dimensional Growth Transition in MoS_2 Chemical Vapor Deposition Synthesis. *Adv. Mater. Interfaces* **2016**, *4*, 1600687. [CrossRef]
28. Splendiani, A.; Sun, L.; Zhang, Y.; Li, T.; Kim, J.; Chim, C.Y.; Galli, G.; Wang, F. Emerging Photoluminescence in Monolayer MoS_2. *Nano Lett.* **2010**, *10*, 1271–1275. [CrossRef] [PubMed]
29. Yu, Y.; Yu, Y.; Cai, Y.; Li, W.; Gurarslan, A.; Peelaers, H.; Aspnes, D.E.; Van de Walle, C.G.; Nguyen, N.V.; Zhang, Y.W.; et al. Exciton-dominated Dielectric Function of Atomically Thin MoS_2 Films. *Sci. Rep.* **2015**, *5*, 16996. [CrossRef] [PubMed]
30. Eilers, P.H.C.; Goeman, J.J. Enhancing scatterplots with smoothed densities. *Bioinformatics* **2004**, *20*, 623–628. [CrossRef] [PubMed]
31. Scheuschner, N.; Ochedowski, O.; Kaulitz, A.M.; Gillen, R.; Schleberger, M.; Maultzsch, J. Photoluminescence of freestanding single- and few-layer MoS_2. *Phys. Rev. B* **2014**, *89*, 125406. [CrossRef]
32. Kataria, S.; Wagner, S.; Cusati, T.; Fortunelli, A.; Iannaccone, G.; Pandey, H.; Fiori, G.; Lemme, M.C. Growth-Induced Strain in Chemical Vapor Deposited Monolayer MoS_2: Experimental and Theoretical Investigation. *arXiv* **2017**, arXiv:1703.00360.
33. Yang, L.; Cui, X.; Zhang, J.; Wang, K.; Shen, M.; Zeng, S.; Dayeh, S.A.; Feng, L.; Xiang, B. Lattice strain effects on the optical properties of MoS_2 nanosheets. *Sci. Rep.* **2014**, *4*. [CrossRef] [PubMed]

![electronics logo] *electronics*

MDPI

Article

Energetic Stabilities, Structural and Electronic Properties of Monolayer Graphene Doped with Boron and Nitrogen Atoms

Seba Sara Varghese [1,2], Sundaram Swaminathan [3,*], Krishna Kumar Singh [4] and Vikas Mittal [2,*]

1   Department of Electrical and Electronics Engineering, Birla Institute of Technology and Science-Pilani, Dubai Campus, Dubai 345055, United Arab Emirates; sebavarghese@gmail.com
2   Department of Chemical Engineering, The Petroleum Institute, Abu Dhabi 2533, United Arab Emirates
3   Department of Electronics and Communication Engineering, DIT University (DITU), Dehradun 248009, India
4   Department of Physics, Birla Institute of Technology and Science-Pilani, Dubai Campus, Dubai 345055, United Arab Emirates; singh@dubai.bits-pilani.ac.in
*   Correspondence: pvc@dituniversity.edu.in (S.S.); vmittal@pi.ac.ae (V.M.); Tel.: +91-135-300-1518 (S.S.); +971-2-607-5491 (V.M.)

Academic Editors: Yoke Khin Yap and Zhixian Zhou
Received: 6 November 2016; Accepted: 6 December 2016; Published: 14 December 2016

Abstract: The structural, energetic, and electronic properties of single-layer graphene doped with boron and nitrogen atoms with varying doping concentrations and configurations have been investigated here via first-principles density functional theory calculations. It was found that the band gap increases with an increase in doping concentration, whereas the energetic stability of the doped systems decreases with an increase in doping concentration. It was observed that both the band gaps and the cohesive energies also depend on the atomic configurations considered for the substitutional dopants. Stability was found to be higher in N-doped graphene systems as compared to B-doped graphene systems. The electronic structures of B- and N-doped graphene systems were also found to be strongly influenced by the positioning of the dopant atoms in the graphene lattice. The systems with dopant atoms at alternate sublattices have been found to have the lowest cohesive energies and therefore form the most stable structures. These results indicate an ability to adjust the band gap as required using B and N atoms according to the choice of the supercell, i.e., the doping density and substitutional dopant sites, which could be useful in the design of graphene-based electronic and optical devices.

Keywords: structural; energetic; electronic; density functional theory; band gap; stability; doped graphene; cohesive energies

1. Introduction

Graphene, a single atomic layer of graphite which exhibits exceptional structural, mechanical, electrical, optical, and chemical properties, has applications in numerous fields [1]. Graphene has attracted the attention of researchers from both experimental and theoretical points of view after its successful isolation in 2004 [2], especially for electronics owing to its exceptional properties [3] such as ballistic electron transport at room temperature [4], high charge carrier mobility [5], room-temperature fractional quantum Hall effect [6], and finite electrical conductivity at zero charge carrier density [7]. These features of graphene that make it a potential candidate for future nanoelectronics [8,9] arise from its unique zero energy band gap with linear energy-momentum relation around the Dirac point [10,11]. However, the absence of a band gap in graphene limits its applications in various nanoelectronic devices, such as *p-n* junction diodes and transistors, and in other energy-related devices such as supercapacitors, solar cells, and fuel cells.

In order to exploit the potential of graphene for electronics, a sizeable band gap should open up in graphene. Until now, various approaches such as application of an external electric field [12], chemical functionalization [13], use of graphene nanoribbons [14], and doping with heteroatoms [15] have been proposed for opening a band gap in graphene. Among these, substitutional doping is suggested to be the most effective method for modifying the electronic properties of graphene, due to the strong dependency of the material properties on the structure. Doping of graphene with other elements would not result in significant degradation of other favorable features of graphene that makes it suitable for enabling nano-sized electronics. The introduction of dopants into the graphene lattice could also lead to important modifications of physical and chemical properties that could be tailored for developing various graphene-based devices with applications in sensing [16–22], energy storage [23–26], gas storage [27–29], etc. Among various dopant atoms, boron (B) and nitrogen (N) atoms have gained significant research attention being the nearest neighbors to carbon (C) that provide a strong probability of entering the graphene lattice and due to the electron acceptor and donor nature of B and N atoms that produces *p*-type and *n*-type graphene, respectively [30–33]. The *p*-type and *n*-type graphene sheets produced by B- and N-doping could be employed for the fabrication of complementary devices in future graphene-based electronic circuits.

There have been many reports on band gap engineering of graphene using substitutional doping [15,30,34–47]. For instance, Wu et al. [37] investigated the geometry, electronic structure, and magnetic properties of graphene doped with light non-metallic atoms such as B, N, O, and F. An ab initio study on the band gap opening in graphene by single B- and N-atom doping in 8, 18, 32, and 50 host C atoms has also been reported [42]. All these works on doped graphene systems have shown that dopant atoms modify the electronic band structure of graphene by introducing an energy gap so that the behavior of graphene changes from semi-metallic to semiconducting. However, one B- and N-atom doping in $3N \times 3N$ (where N is an integer) graphene supercells have shown zero band gap at the Dirac point [38,42,44], whereas, in the case of the one B- and N-atom doping in the $(3N - 1) \times (3N - 1)$ and $(3N + 1) \times (3N + 1)$ supercells of pristine graphene, there is a band gap which can be tunable by the doping concentration. Zhou et al. [44] discovered an interesting 3N rule for periodically doped graphene sheets, which suggests that when the primitive cell is $3N \times 3N$, the doped graphene has a zero gap or negligible gap, and the properties of doped graphene can be predicted by their primitive cell sizes.

The effect of doping graphene with B and N concentrations varying from 2% to 12% (simulated by varying the number of dopants from one to six in 50 host atoms) on the geometry and electronic structure of single-layer graphene has been systematically analyzed by Rani and Jindal [30]. They observed a dependence of the band gap not only on the concentration of dopants, but also on the position of the dopant atom in the graphene sheet. The results showed a maximum band gap upon placing the dopants at the same sublattice locations and a minimum band gap upon placing the dopants at alternate sublattice locations of the graphene. Another study presented the electronic and magnetic properties of single-layer graphene doped with N atoms and analyzed the dependence of magnetic moments and band gaps in graphene on N-substitutional doping configurations by considering two N atoms in graphene supercells containing 8, 18, and 32 host C atoms [43].

A systematic analysis of the structural and electronic properties of N-doped graphene with two N-substitutional dopants in $3N \times 3N$ graphene supercells by considering different doping configurations has not been reported. To the best of our knowledge, a similar study on B-doped graphene with more than one dopant in $3N \times 3N$ graphene supercells has also still not appeared in the literature. In this paper, we investigate the atomic structures, the stabilities, and the electronic properties—specifically the band structures of graphene doped with B and N atoms in 8, 18, 32 and 72 host C atoms. As B and N atoms can be placed at C sites of the crystal lattice in many different configurations, several substitutional dopant sites in the graphene sheet are analyzed. The effect of

B- and N-doping on the structural and electronic properties of graphene is analyzed by varying the doping concentrations from 1.39% to 25% and by considering different configurations for the same doping concentration. The dependence of the cohesive energy per atom on the doping concentration and the different doping configurations are also studied to understand the stabilities and to compare the energetics of the B- and N-doped systems. Fourteen doping concentrations between 1.39% and 25% are considered for the study.

2. Computational Method

All calculations are performed within the framework of density functional theory (DFT) as implemented in ABINIT code [48]. The generalized gradient approximation (GGA) exchange-correlation (XC) functional in the Perdew–Burke–Ernzerhof (PBE) form [49] is adopted in the structural optimization and electronic structure calculations of both pristine graphene and different doped graphenes. Norm-conserving Troullier–Martins type pseudopotentials [50] are used to describe the electron–ion interactions. The energy convergence criterion is chosen to be ~10 meV/atom. A plane-wave basis set with converged cutoff energy of 816 eV is used (see Supplementary Figure S1). The sampling of the Brillouin zone is performed using the k-point mesh generated by the Monkhorst–Pack scheme [51]. Converged k-point grids corresponding to a $24 \times 24 \times 1$ grid for a graphene unit cell are used for different graphene supercells (see Supplementary Figure S2). For all systems, the relaxation of basis vectors and atomic coordinates are performed by minimizing the total energy. Structural optimization has been conducted using the Broyden–Fletcher–Goldfard–Shanno (BFGS) minimization until the residual forces on atoms are lower than 0.0025 eV/Å.

A single-layer graphene sheet is modeled using four different supercell sizes, i.e., a 2×2 $((3N - 1) \times (3N - 1)$, where $N = 1)$ supercell with 8 C atoms, a 3×3 $(3N \times 3N$, where $N = 1)$ supercell with 18 C atoms, a 4×4 $((3N + 1) \times (3N + 1)$, where $N = 1)$ supercell with 32 C atoms, and a 6×6 $(3N \times 3N$, where $N = 2)$ supercell with 72 C atoms, where the distance between the adjacent graphene layers along the perpendicular direction is taken as 10 Å to avoid interlayer interactions due to periodic boundary conditions. B- and N-doped graphenes are simulated by replacing the C atom in the supercell structure by a B or N atom and by choosing the corresponding pseudopotentials. B- and N-doping concentrations from 1.39% to 25% are modeled through the substitution of one and two C atoms in the 2×2 supercell by dopant atoms, which corresponds to 12.5% and 25% doping concentrations, respectively; the substitution of one, two, three, and four C atoms in the 3×3 supercell by dopant atoms, which corresponds to 5.56% 11.11%, 16.67%, and 22.22% doping concentrations, respectively; the substitution of one and two C atoms in the 6×6 supercell by dopant atoms, which corresponds to 1.39% and 2.78% doping concentrations, respectively; the substitution of one, two, three, four, five, and six C atoms in the 4×4 supercell by dopant atoms, which corresponds to 3.13%, 6.25%, 9.38%, 12.5%, 15.63%, and 18.75% doping concentrations, respectively.

In all cases, we first optimize the geometry of the B- and N-doped systems. The cohesive energy per atom, E_{coh}, is calculated as [30]

$$E_{coh} = \frac{(E_{tot} - n_i E_i)}{n}, i = C, B, N$$

where E_{tot} and E_i represent the total energies of the considered doped system and of the individual elements present within the doped system. The total energies of the individual elements (C, B, or N) are calculated by defining a large supercell and adding the element (C, B, or N) at $(0, 0, 0)$. n is the total number of atoms present in the system, and n_i is the total number of species i present in the configuration. The values of E_{coh} indicate the energetic stability of the systems. The lesser the value, the more stable the system is. Finally, the electronic band structures are computed for the optimized doped systems from which the widths of the band gap are determined. We elucidate the dependence of the band gap and the cohesive energy per atom on the concentration and position of the dopant atoms.

3. Results

3.1. Pristine Graphene

Upon structural optimization of pristine graphene (PG), the lattice constant and the C–C bond length were observed to be 2.458 Å and 1.42 Å. The calculated lattice constant is in good agreement with the experimental value of 2.46 Å [52], and the calculated C–C bond length of 1.42 Å as seen in Figure 1a–d is very close to the experimental value of 1.421 Å [3,52]. The relaxed geometries and band structures of 2 × 2, 3 × 3, 4 × 4, and 6 × 6 graphene supercells obtained from the calculations are shown in Figures 1a–d and 2a–d, respectively.

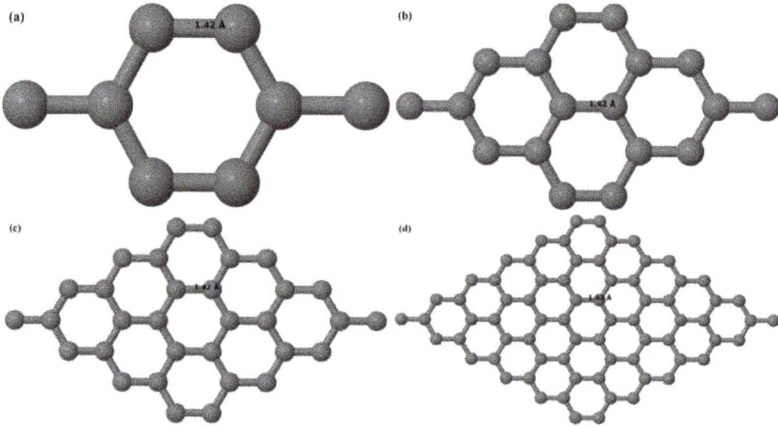

Figure 1. Optimized structures of (**a**) 2 × 2; (**b**) 3 × 3; (**c**) 4 × 4; (**d**) 6 × 6 supercells of pristine graphene (PG), the carbon atoms are shown in gray.

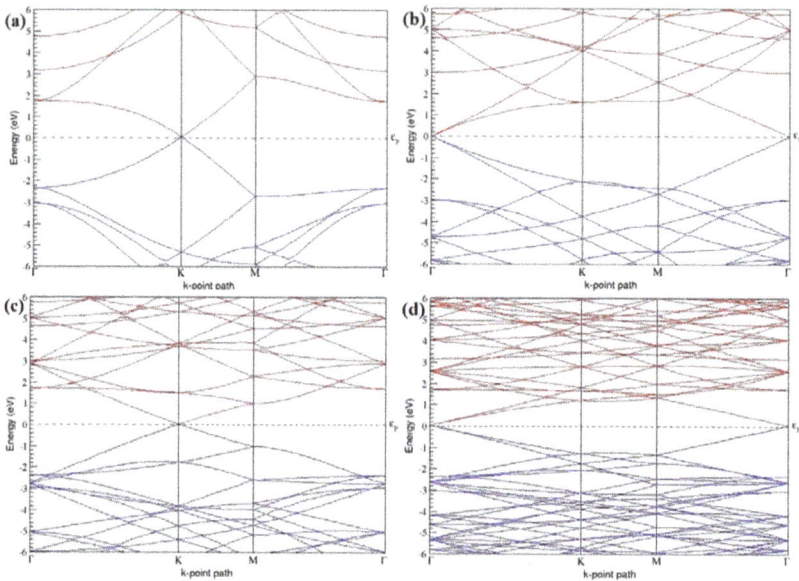

Figure 2. Band structures of (**a**) 2 × 2; (**b**) 3 × 3; (**c**) 4 × 4; (**d**) 6 × 6 supercells of PG.

The band structures of the PG sheets (Figure 2a–d) along the high symmetry points (Γ-K-M-Γ) of the hexagonal Brillouin zone of graphene, which exhibit zero energy band gap and linear in-plane dispersion around the Fermi level, are found to be in good agreement with the reported literature [7], which presents the reliability of the employed calculation method. In the band structures of 2 × 2 and 4 × 4 graphene supercells, the top of the valence band and the bottom of the conduction band degenerate at the K-point as seen in Figure 2a,c, whereas, in 3 × 3 and 6 × 6 supercells, the degeneracy is observed at the Γ-point, as presented in Figure 2b,d. This is in accordance with the previous reports that the Dirac point moves into the Γ-point, when the supercells are dimensions of three (3 × 3, 6 × 6, 9 × 9 supercells, etc.) [44].

After successful reproduction of the structural and the electronic band structures of PG, PG is doped with 14 different concentrations of B and N atoms. The study of the change in the geometries and the electronic structures of graphene upon doping with varying B- and N-atom concentrations and the analysis of the band gap for each doping concentration and for different dopant sites for the same doping concentration are carried out as described below.

3.2. B-Doped Graphene

3.2.1. B-Doped Graphene System with One B Atom per Supercell

Here, we consider one B substitutional dopant in 2 × 2, 3 × 3, 4 × 4, and 6 × 6 graphene supercells. Upon structural optimization of all graphene supercells doped with one B atom, it was observed that the planar geometry of PG remains undisturbed (Figure 3a–d) even after the introduction of B atom, as B also undergoes sp^2 hybridization like the other C atoms in the crystal lattice, which is in accordance with earlier results [30,37]. The optimized lattice constant increases from 2.458 Å to 2.464 Å, 2.471 Å, 2.482 Å, and 2.514 Å for 6 × 6, 4 × 4, 3 × 3, and 2 × 2 supercells doped with one B atom (1.39%, 3.13%, 5.56%, 12.5% B concentrations), respectively. Since the atomic radius of B is larger than that of C, the lattice constant increases with the increase in the B-doping concentration, showing agreement with previous reports [30].

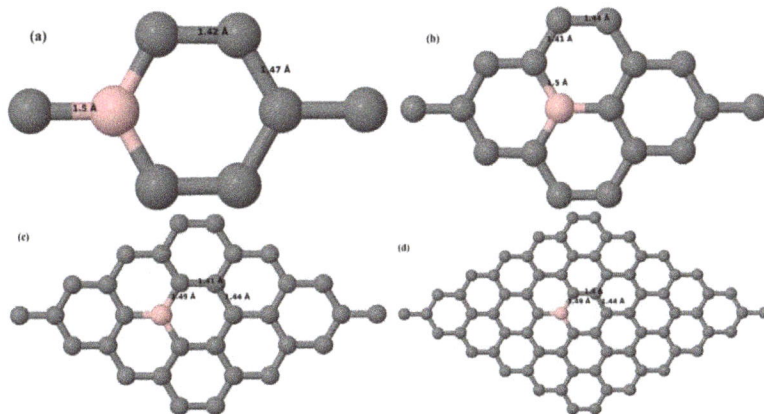

Figure 3. Optimized structures of various graphene systems doped with one B atom, the B atoms are shown in a rose color; (**a**) 2 × 2 graphene supercell with 12.5% B concentration; (**b**) 3 × 3 graphene supercell with 5.56% B concentration; (**c**) 4 × 4 graphene supercell with 3.13% B concentration; (**d**) 6 × 6 graphene supercell with 1.39% B concentration.

Figure 3a–d depict the optimized geometries of 2 × 2, 3 × 3, 4 × 4, and 6 × 6 graphene supercells doped with one B atom. The large covalent radius of B compared to that of C results in the expansion of the C–B bond length [30,37] to 1.5 Å for one B doping in 2 × 2 and 3 × 3 supercells (Figure 3a,b), whereas, in 4 × 4 and 6 × 6 supercells, the C–B bond length was extended to 1.49 Å (Figure 3c,d) [30,37] from the ideal C–C bond length of 1.42 Å. The C–C bond lengths adjacent to the B atom were reduced from 1.42 Å to 1.41 Å for 3 × 3 and 4 × 4 graphene supercells (Figure 3b,c), whereas, in the case of the 6 × 6 graphene supercell, it was reduced to 1.4 Å (Figure 3d) in order to compensate for the long C–B bond in the crystal structure so as to retain the planar geometry. The observed reduction in the C–C bond length in the close proximity to the B atom is in agreement with the results reported in [30]. The calculated cohesive energies per atom of the relaxed geometries of graphene systems doped with one B atom are presented in Table 1.

Figure 4a–d present the band structures computed for the optimized structures of different graphene systems doped with one B atom shown in Figure 3a–d. Since the planar geometry of graphene is well preserved even after one B doping, the linear energy dispersion remains unaltered along the high symmetry points of the Brillouin zone as seen in Figure 4a–d, similar to the reported literature [30,37]. Due to the symmetry breaking of the graphene sublattices by the introduction of the B atom, the band structures of 2 × 2 and 4 × 4 graphene supercells show a band gap of ~0.66 eV (Figure 4a) and ~0.19 eV (Figure 4c) at the Dirac point for one B-atom doping (corresponding to 12.5% and 3.13% B concentrations), respectively. The present results are slightly greater that the values reported in [42], probably due to the variation in the employed computational method. The 3 × 3 and 6 × 6 graphene supercells doped with one B atom (corresponding to 5.56% and 1.39% B concentrations) do not show any band gap (Figure 4b,d), which is found to be in agreement with the reported zero band gap phenomenon in 3N × 3N graphene supercells [38,42,44]. In general, the energy gap increases from ~0.19 eV to ~0.66 eV for 4 × 4 and 2 × 2 supercells doped with one B atom (3.13% and 12.5% B concentrations), respectively (Table 1).

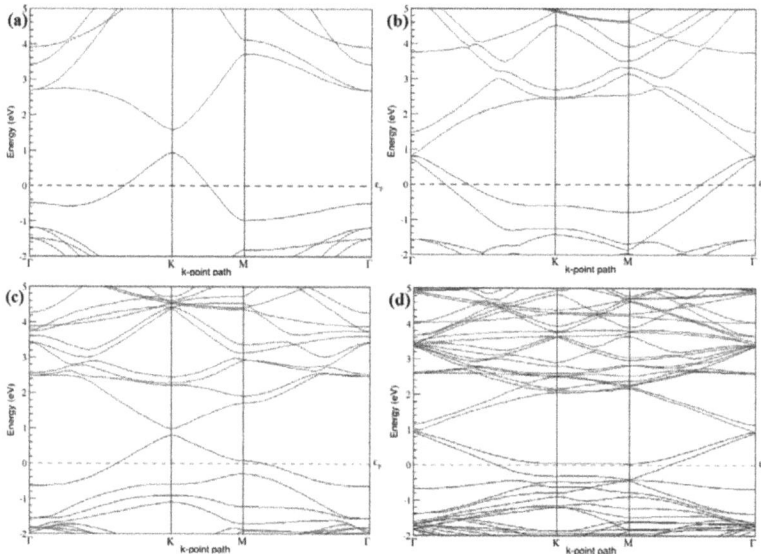

Figure 4. Band structures of graphene systems doped with one B atom corresponding to the optimized structures shown in Figure 3a–d; (**a**) 2 × 2 graphene supercell with 12.5% B concentration; (**b**) 3 × 3 graphene supercell with 5.56% B concentration; (**c**) 4 × 4 graphene supercell with 3.13% B concentration; (**d**) 6 × 6 graphene supercell with 1.39% B concentration.

Both 2 × 2 and 4 × 4 graphene supercells doped with one B atom exhibit *p*-type doping electronic properties with band gaps, whereas the 3 × 3 and 6 × 6 graphene supercells doped with one B atom show *p*-type doping properties with zero band gap.

Table 1. The B concentrations, cohesive energies, and the band gap introduced for various supercells doped with one B atom.

Model	B Concentration (%)	E_{coh} (eV/atom)	Band Gap (eV)
2 × 2	12.5	−8.795	0.658
3 × 3	5.56	−9.088	0
4 × 4	3.13	−9.197	0.190
6 × 6	1.39	−9.270	0

3.2.2. B-Doped Graphene System with Two B Atoms per Supercell

As graphene's honeycomb lattice consists of two interpenetrating triangular sublattices as shown in Figure 5, several isomers of the same doping concentration are possible. A few isomers with configurations of three doping sites, i.e., when all the dopant atoms are adjacent, when they are at same sublattice positions (all in either sublattice "A" or in sublattice "B"), and when they are at different sublattice positions (in sublattice "A" and "B"), are only presented here for simplicity, as all possible doping configurations of any atomic doping concentration will fall under these three categories only. Hence, the cohesive energies and the band structures of the above-mentioned geometries corresponding to the same doping concentration are calculated to analyze the influence of the dopant sites on the stabilities and the band gap values.

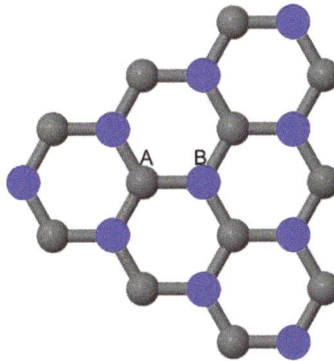

Figure 5. Schematic illustration of the honeycomb lattice of graphene. Atoms in sublattices A and B points are shown in gray and blue color respectively.

Here we consider the substitution of two C atoms by two B atoms in 2 × 2, 3 × 3, 4 × 4, and 6 × 6 graphene supercells (Figure 6a–l). Similar to graphene systems doped with one B atom, the graphene systems doped with two B atoms also retain the planar geometry of PG as seen in Figure 6a–l. The optimized lattice constant increases from 2.458 Å to 2.469 Å, 2.484 Å, 2.504 Å, and 2.576 Å for 6 × 6, 4 × 4, 3 × 3, and 2 × 2 supercells doped with two B atoms (2.78%, 6.25%, 11.11%, and 25% B concentrations), respectively, which also shows an increase in lattice constant with increasing B-doping concentration similar to that observed for graphene systems doped with one B atom.

Figure 6a–l depict the optimized geometries of the 2 × 2, 3 × 3, 4 × 4, and 6 × 6 graphene supercells doped with two B atoms. Three doping configurations of the 2 × 2, 3 × 3, 4 × 4, and 6 × 6 supercells doped with two B atoms (corresponding to 25%, 11.11%, 6.25%, and 2.78% B concentrations, respectively) are considered with the dopant atoms at adjacent, (Figure 6a,d,g,j), same (Figure 6b,e,h,k), and alternate sublattice points (Figure 6c,f,i,l). The C–B bond length was expanded significantly (Figure 6a–l) as compared to graphene systems doped with one B atom, due to the presence of the two B atoms, which have a bigger size compared to the other C atoms in the lattice, whereas the C–C bond length in close proximity of the dopant was shortened from 1.42 Å to 1.41 Å or 1.4 Å, or even to 1.39 Å, in most of the optimized structures, based on the position of the B dopants, in an attempt to the preserve the planar lattice structure. After obtaining stable geometries, the cohesive energies were calculated for all the considered graphene systems doped with two B atoms and are listed in Table 2.

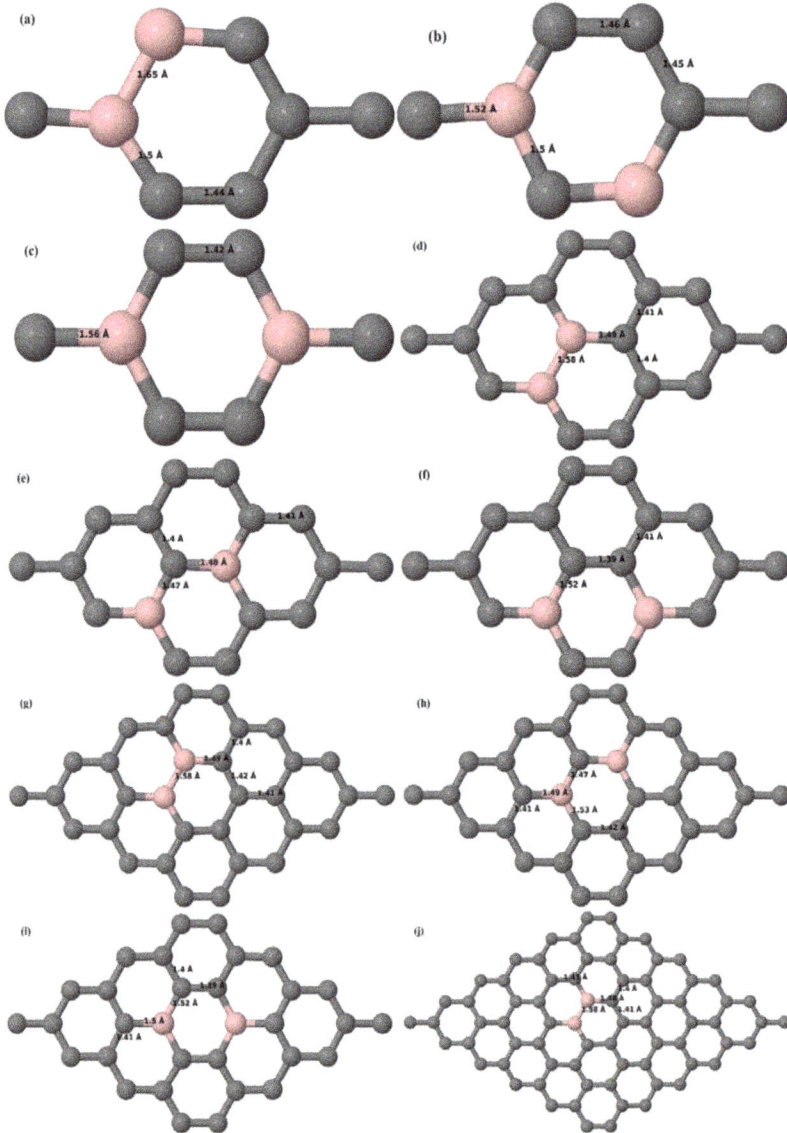

Figure 6. *Cont.*

(k)

(l)

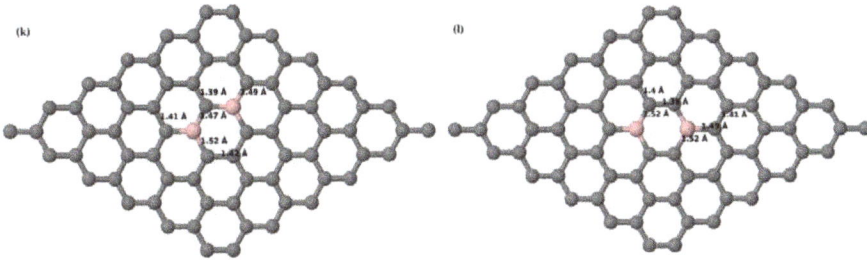

Figure 6. Optimized structures of various graphene systems doped with two B atoms with different doping configurations; (**a–c**) 2 × 2 graphene supercell with 25% B concentration; (**d–f**) 3 × 3 graphene supercell with 11.11% B concentration; (**g–i**) 4 × 4 graphene supercell with 6.25% B concentration; (**j–l**) 6 × 6 graphene supercell with 2.78% B concentration.

Table 2. The B concentrations, doping configurations with considered sublattice, cohesive energy, and the band gap introduced for various supercells doped with two B atoms.

Model	B Concentration (%)	Configuration	Considered Sublattices for Dopants	E_{coh} (eV/atom)	Band Gap (eV)
2 × 2	25	Figure 6a	"B" and "A" (adjacent)	−8.284	0.492
		Figure 6b	Both in "B" (same)	−8.353	1.282
		Figure 6c	"B" and "A" (alternate)	−8.493	0
3 × 3	11.11	Figure 6d	"B" and "A" (adjacent)	−8.842	0.265
		Figure 6e	Both in "B" (same)	−8.874	0.274
		Figure 6f	"B" and "A" (alternate)	−8.921	0.046
4 × 4	6.25	Figure 6g	"B" and "A" (adjacent)	−9.058	0.130
		Figure 6h	Both in "B" (same)	−9.079	0.375
		Figure 6i	"B" and "A" (alternate)	−9.095	0.047
6 × 6	2.78	Figure 6j	"B" and "A" (adjacent)	−9.202	0.100
		Figure 6k	Both in "B" (same)	−9.212	0.114
		Figure 6l	"B" and "A" (alternate)	−9.219	0.008

Figure 7a–l present the band structures computed for the optimized structures of different B-doped graphene systems shown in Figure 6a–l. As seen in Figure 7a–l, the linear dispersion around the Dirac point is not completely destroyed, but an energy band gap opens in all cases except for graphene doped with two B atoms into the 2 × 2 graphene supercell with dopant atoms at the alternate sublattice points (Figure 7c). The 2 × 2 graphene supercell doped with two B atoms (corresponding to 25% B concentration) shows band gaps of ~0.49 eV (Figure 7a) and ~1.28 eV (Figure 7b) upon placing the B atoms at adjacent and same sublattice positions, respectively, in graphene. B-doped graphene with 25% B concentration has a zero gap (Figure 7c) when the B atoms are at alternate sublattice positions due to the symmetry formed by the B atoms situated in two graphene sublattices ("A" and "B"). At a 6.25% B concentration, band gaps of 0.13 eV (Figure 7g), 0.375 eV (Figure 7h), and ~0.05 eV (Figure 7i) open up in graphene when the B atoms are placed at the adjacent, same, and alternate sublattice positions in graphene, respectively.

The band structures of 3 × 3 and 6 × 6 graphene supercells doped with two B atoms are characterized by a non-zero band gap (Figure 7d–f,j–l), which is different from the zero band gap behavior observed for the 3 × 3 and 6 × 6 graphene supercells doped with one B atom. The 3 × 3 graphene supercells doped with two B atoms (corresponding to 11.11% B concentration) exhibit band gaps of 0.265 eV (Figure 7d), 0.274 eV (Figure 7e), and ~0.05 eV (Figure 7f) for the configurations with dopants at adjacent, same, and alternate sublattices in graphene, respectively. Doping of graphene with 2.78% concentration of B results in a band gap opening of 0.10 eV (Figure 7j), ~0.11 eV (Figure 7k), and 0.008 eV (Figure 7l), respectively, in graphene upon placing the B atoms at adjacent, same, and alternate sublattice sites in graphene.

The $2 \times 2, 3 \times 3, 4 \times 4$, and 6×6 graphene supercells doped with two B atoms exhibited *p*-type semiconducting behaviors. The 2×2 graphene supercell doped with two B atoms also exhibit *p*-type electronic behavior, but does not show any band gap when the dopants are at alternate sublattice sites (Figure 7c). The considered B concentrations, the doping configurations along with the selected sublattices, the calculated cohesive energies, and the band gaps observed for all considered graphene systems doped with two B atoms are summarized in Table 2.

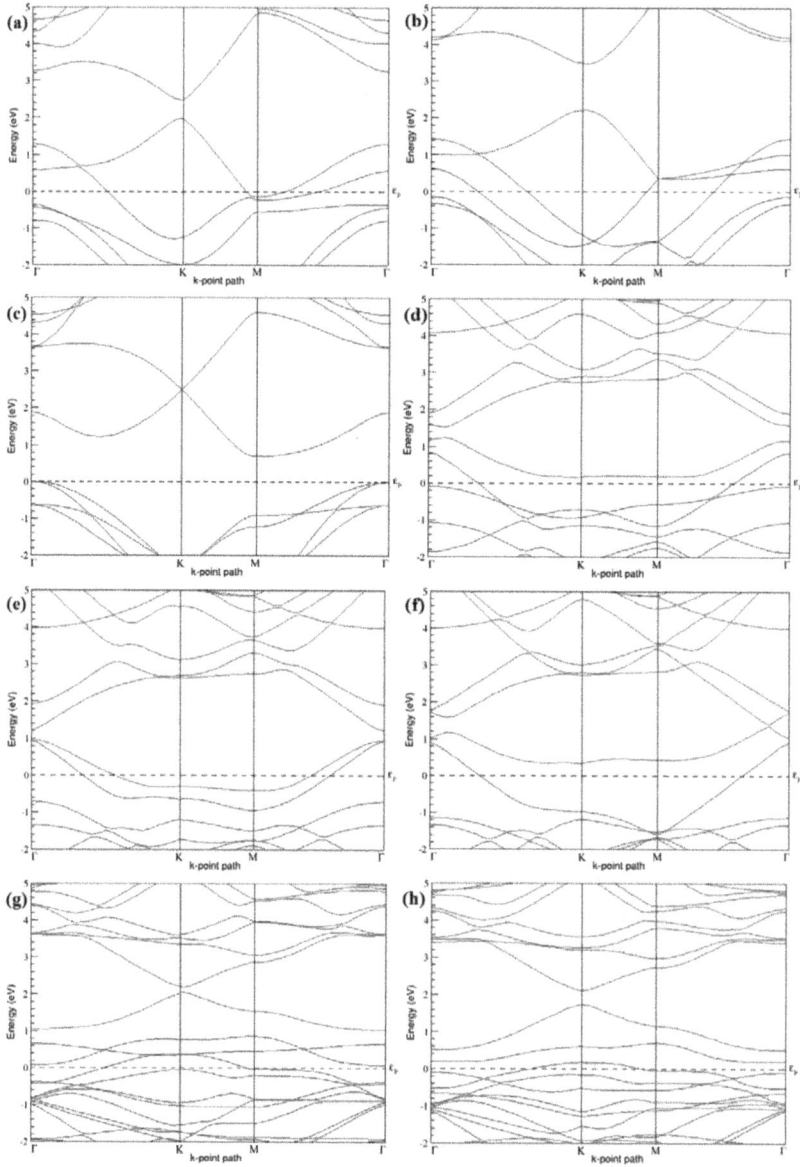

Figure 7. *Cont.*

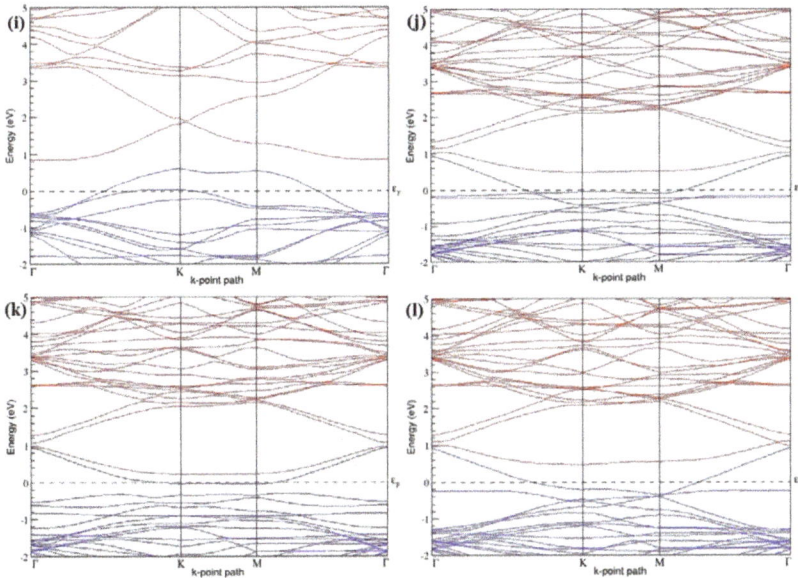

Figure 7. Band structures of graphene systems doped with two B atoms corresponding to the optimized structures shown in Figure 6a–l; (**a–c**) for 2 × 2 graphene supercell with 25% B concentration; (**d–f**) 3 × 3 graphene supercell with 11.11% B concentration; (**g–i**) 4 × 4 graphene supercell with 6.25% B concentration; (**j–l**) 6 × 6 graphene supercell with 2.78% B concentration.

3.2.3. B-Doped Graphene System with Three B Atoms per Supercell

Here we consider the substitution of three C atoms by three B atoms in 3 × 3 and 4 × 4 graphene supercells. Similar to graphene systems doped with one and two B atoms, the planar lattice structure of graphene remains the same (Figure 8a–f) even after the introduction of three B atoms. The optimized lattice constant increases from 2.458 Å to 2.475 Å and 2.528 Å for 4 × 4 and 3 × 3 supercells doped with three B atoms (9.38% and 16.67% B concentrations), respectively, which indicates increase in lattice constant with doping concentration similar to that seen in graphene systems doped with one and two B atoms.

Figure 8a–f present the relaxed geometries of 3 × 3 and 4 × 4 graphene supercells doped with three B atoms at adjacent, same, and alternate sublattices, respectively. After structural optimization, it was found that graphene structures doped with three B atoms at adjacent locations experience significant geometrical distortion (Figure 8a,d) due to the positioning of three B atoms in the same six-membered carbon ring, as compared to the other graphene systems doped with three B atoms. However, the three adjacent B atoms are still seen to lie within the plane with large adjustments in the adjoining bond lengths (Figure 8a,d).

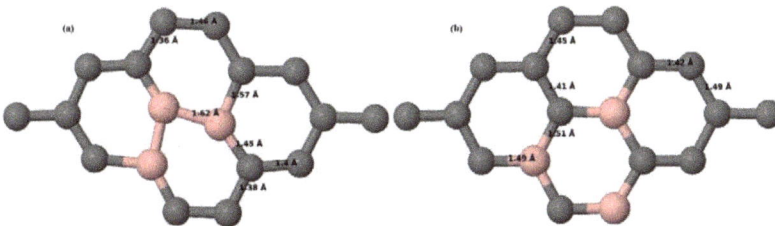

Figure 8. *Cont.*

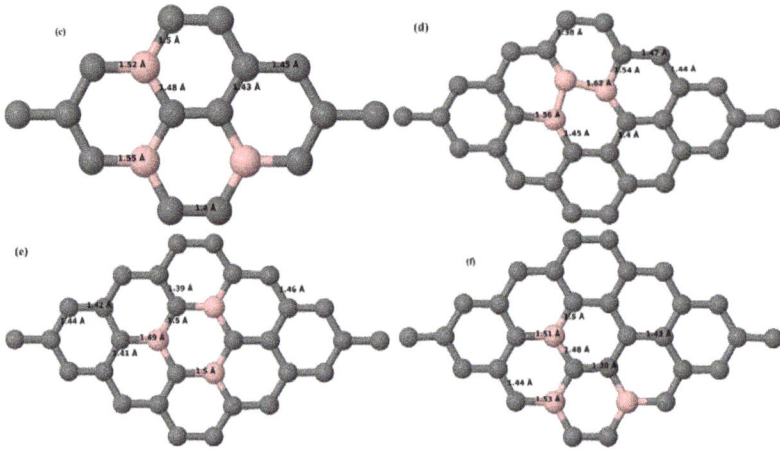

Figure 8. Optimized structures of various graphene systems doped with three B atoms with different doping configurations; (**a–c**) 3 × 3 graphene supercell with 16.67% B concentration; (**d–f**) 4 × 4 graphene supercell with 9.38% B concentration.

Figure 9a–f present the band structures computed for the optimized structures of different graphene systems doped with three B atoms shown in Figure 8a–f. The band structures of both 3 × 3 and 4 × 4 graphene supercells doped with three B atoms show non-zero band gaps as seen in Figure 9. The linear energy dispersion at the Dirac point in the band structures of graphene systems doped with three B atoms is found to be greatly affected (Figure 9a,d) for those structures in which the B-substitutional dopants are placed at the adjacent positions (Figure 8a,d), which could be attributed to their highly distorted geometries. The observation of highly deformed band structures of B-doped graphene with an odd number of dopants in adjacent positions is consistent with that reported in [30].

At a 16.67% B concentration, the observed band gaps are ~0.03 eV (Figure 9a) and ~0.11 eV (Figure 9c), when the dopant atoms are placed at the adjacent and alternate sublattices, respectively. The positioning of the three B atoms at the same sublattices leads to a large band gap opening of ~0.90 eV (Figure 9b) at a 16.67% B concentration. Graphene with a 9.38% B concentration shows band gaps of 0.235 eV (Figure 9d), ~0.57 eV (Figure 9e), and ~0.16 eV (Figure 9f), for the doping configurations of adjacent, same, and alternate sublattices, respectively.

Graphene with three B-doping atoms in 3 × 3 and 4 × 4 graphene supercells exhibit *p*-type semiconducting behaviors. Table 3 presents the models used, the B concentrations, the considered doping configurations with sublattices, the calculated cohesive energies, and the band gaps introduced for all graphene systems doped with three B atoms.

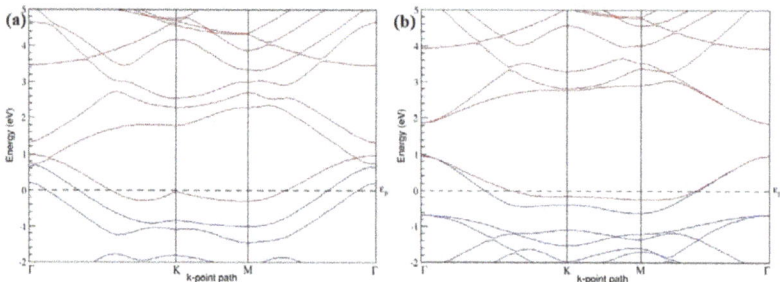

Figure 9. *Cont.*

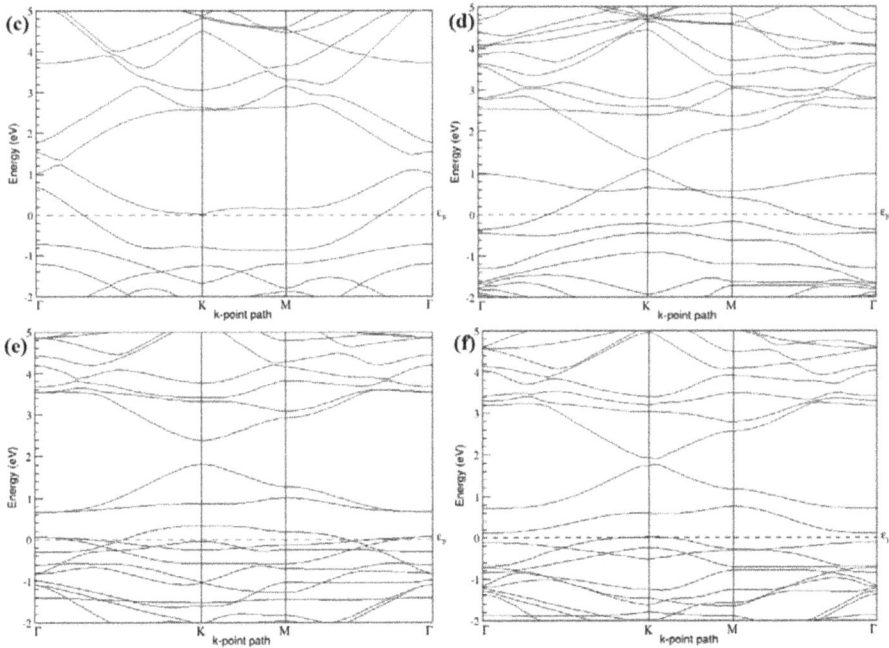

Figure 9. Band structures of graphene systems doped with three B atoms corresponding to the optimized structures shown in Figure 8a–f; (**a–c**) for 3 × 3 graphene supercell with 16.67% B concentration; (**d–f**) 4 × 4 graphene supercell with 9.38% B concentration.

Table 3. The B concentrations, doping configurations with considered sublattice, cohesive energies, and the band gaps introduced for various supercells doped with three B atoms.

Model	B Concentration (%)	Configuration	Considered Sublattices for Dopants	E_{coh} (eV/atom)	Band Gap (eV)
3 × 3	16.67	Figure 8a	"B", "A" and "B" (adjacent)	−8.587	0.032
		Figure 8b	All in "B" (same)	−8.658	0.900
		Figure 8c	"B", "B" and "A" (alternate)	−8.688	0.107
4 × 4	9.38	Figure 8d	"B", "A" and "B" (adjacent)	−8.914	0.235
		Figure 8e	All in "B" (same)	−8.952	0.568
		Figure 8f	"B", "B" and "A" (alternate)	−8.966	0.156

3.2.4. B-Doped Graphene System with Four B Atoms per Supercell

Here we consider the substitution of four C atoms by four B atoms in 3 × 3 and 4 × 4 graphene supercells. After structural relaxation, all graphene systems doped with four B atoms appear to have the same planar configuration of PG by adjusting the associated bond lengths (Figure 10a–f). The relaxed lattice constant increases from 2.458 Å to 2.512 Å and 2.557 Å for 4 × 4 and 3 × 3 supercells doped with four B atoms (corresponding to 12.5% and 22.22% B concentrations), respectively, similar to that observed in other B-doped graphene systems.

Figure 10a–f present the relaxed structures of 3 × 3 and 4 × 4 supercells doped with four B atoms at adjacent, same, and alternate sublattices, respectively. As compared with graphene systems doped with three B atoms, graphene systems doped with four B atoms experience much less structural distortion when the B atoms are placed at adjacent positions in the lattice (Figure 10a,d).

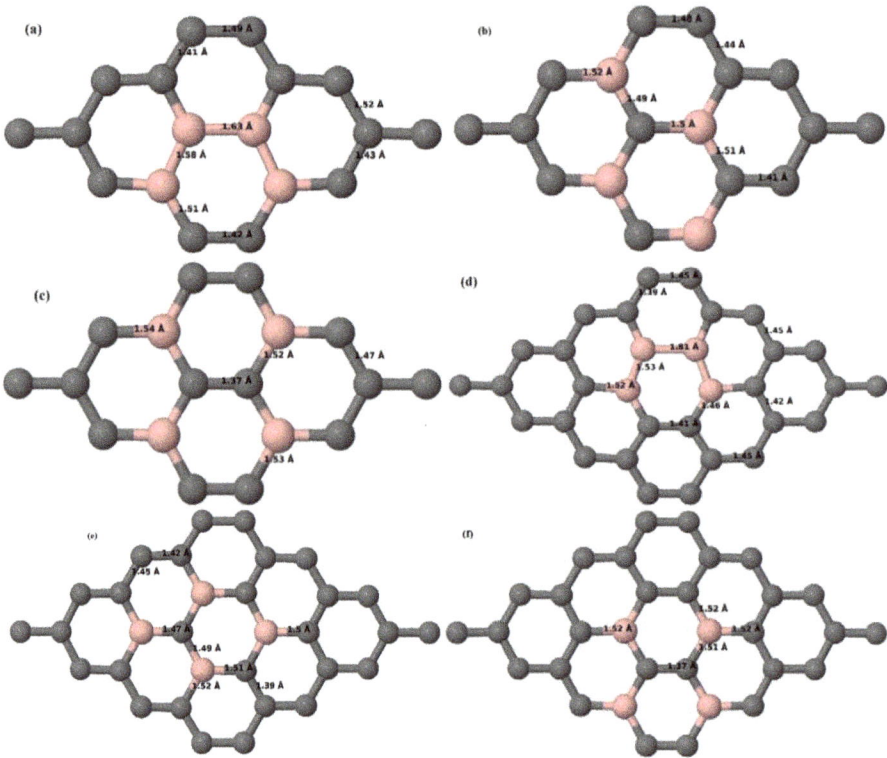

Figure 10. Optimized structures of various graphene systems doped with four B atoms with different doping configurations; (**a–c**) 3 × 3 graphene supercell with 22.22% B concentration; (**d–f**) 4 × 4 graphene supercell with 12.5% B concentration.

Figure 11a–f present the band structures computed for the optimized structures of different graphene systems doped with four B atoms shown in Figure 10a–f. The band structures of both 3 × 3 and 4 × 4 graphene supercells doped with four B atoms depicted in Figure 11a–f show non-zero band gap values. At a 22.22% B concentration, the doped graphene exhibits band gaps of 0.096 eV (Figure 11a), ~0.73 eV (Figure 11b), and ~0.06 eV (Figure 11c) when B atoms are at adjacent, same, and alternate sublattices, respectively. Graphene with 12.5% B concentration shows band gaps of ~0.12 eV (Figure 11d) and ~0.05 eV (Figure 11f) for the doping configurations of adjacent and alternate sublattices, respectively. A maximum band gap of ~0.66 eV (Figure 11e) opens up in graphene at a 12.5% B concentration, when the dopant atoms are at the same sublattice.

All graphene structures doped with four B atoms exhibit *p*-type semiconducting behaviors. Table 4 presents the models used, the doping concentrations with selected sublattices, calculated cohesive energies, and the band gaps introduced for all graphene systems doped with four B atoms.

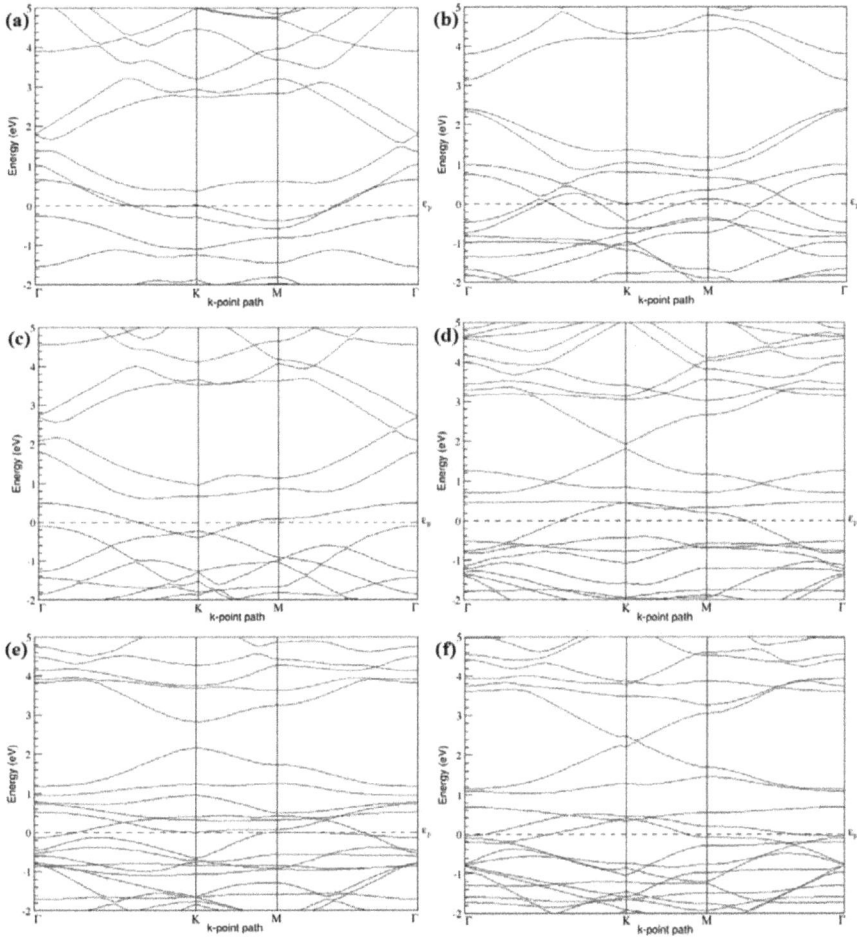

Figure 11. Band structures of graphene systems doped with four B atoms corresponding to the optimized structures shown in Figure 10a–f; (**a–c**) for 3 × 3 graphene supercell with 22.22% B concentration; (**d–f**) 4 × 4 graphene supercell with 12.5% B concentration.

Table 4. The B concentrations, doping configurations with considered sublattice, cohesive energies, and the band gaps introduced for various supercells doped with four B atoms.

Model	B Concentration (%)	Configuration	Considered Sublattices for Dopants	E_{coh} (eV/atom)	Band Gap (eV)
		Figure 10a	"B", "A", "B" and "A" (adjacent)	−8.364	0.096
3 × 3	22.22	Figure 10b	All in "B" (same)	−8.455	0.728
		Figure 10c	"B", "B", "A" and "A" (alternate)	−8.508	0.062
		Figure 10d	"B", "A", "B" and "A" (adjacent)	−8.773	0.118
4 × 4	12.5	Figure 10e	All in "A" (same)	−8.829	0.662
		Figure 10f	"B", "B", "A" and "A" (alternate)	−8.856	0.050

3.2.5. B-Doped Graphene System with Five B Atoms per Supercell

Here we consider the substitution of five C atoms by five B atoms in a 4 × 4 graphene supercell. After structural relaxation, it was observed that all graphene systems doped with five B atoms exhibit planar geometry (Figure 12a–c), similar to other B-doped graphene systems. The relaxed lattice constant increases from 2.458 Å to 2.527 Å for the 4 × 4 graphene supercell doped with five B atoms (corresponding to 15.63% B concentration), as observed in other B-doped graphene systems. Similar to graphene systems doped with three B atoms, graphene systems doped with five B atoms at adjacent positions experience significant structural distortion, but the planar configuration is maintained by adjusting the C–B and adjacent C–C bond lengths (Figure 12a). The observed structural distortion is larger than that observed in systems doped with four B atoms.

The band structures computed for the optimized structures of different graphene systems doped with five B atoms shown in Figure 13a–c, show *p*-type semiconducting nature. The observations of disturbed linear energy dispersion at the Dirac point and highly deformed band structure (Figure 13a), for the doping configuration with five dopants at adjacent positions, is in agreement with similar previous reports [30].

Table 5 summarizes the observed band gaps and the calculated cohesive energies for different doping configurations corresponding to 15.63% B concentration in graphene. At a 15.63% B concentration, B atoms located at adjacent, same, and alternate sublattices open band gaps of 0.245 eV (Figure 13a), ~0.76 eV (Figure 13b), and ~0.13 eV (Figure 13c), respectively, in graphene.

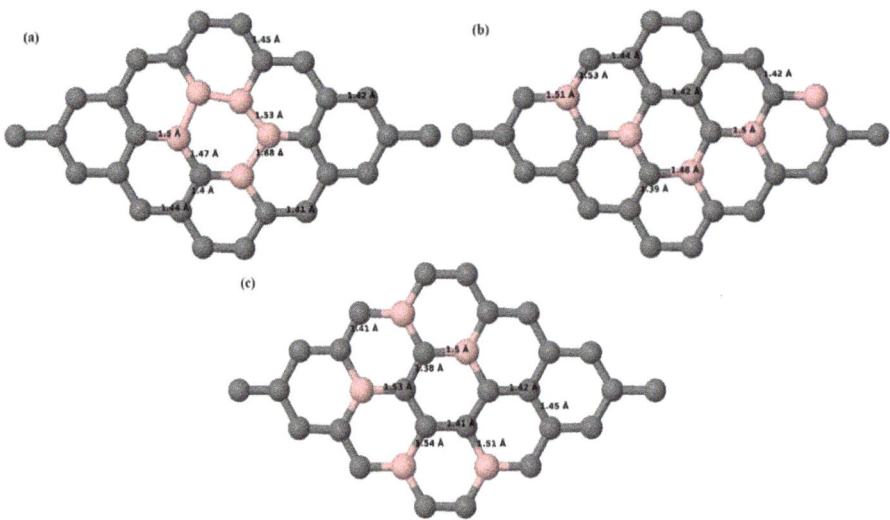

Figure 12. (**a–c**) Optimized structures of the 4 × 4 graphene supercell doped with five B atoms (15.63% B concentration) with different doping configurations.

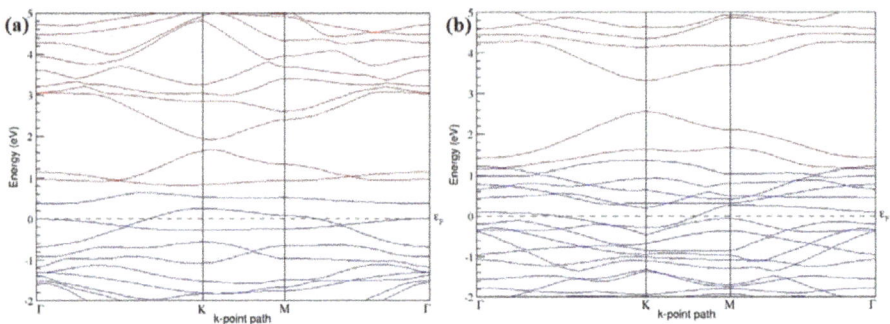

Figure 13. *Cont.*

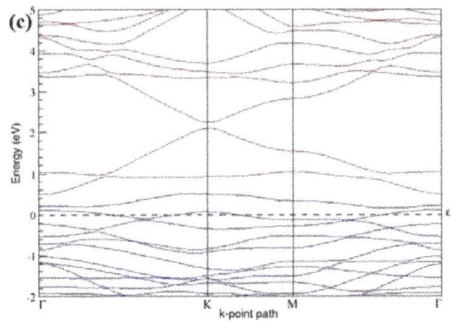

Figure 13. (a–c) Band structures of the 4 × 4 graphene supercell doped with five B atoms (15.63% B concentration) corresponding to the optimized structures shown in Figure 12a–c.

Table 5. The B concentration, doping configurations with considered sublattice, cohesive energies, and the band gaps introduced for 4 × 4 supercell doped with five B atoms.

Model	B Concentration (%)	Configuration	Considered Sublattices for Dopants	E_{coh} (eV/atom)	Band Gap (eV)
		Figure 12a	"B", "A", "B", "A" and "B" (adjacent)	−8.643	0.245
4 × 4	15.63	Figure 12b	All in "B" (same)	−8.714	0.756
		Figure 12c	"B", "A", "B", "A" and "B" (alternate)	−8.746	0.133

3.2.6. B-Doped Graphene System with Six B Atoms per Supercell

Here we consider the substitution of six C atoms in a 4 × 4 graphene supercell by six B atoms in which the dopant positions at adjacent, same, and alternate sublattices are considered. The planar structure is preserved even after the introduction of six B atoms in the lattice (Figure 14a–c). The relaxed lattice constant increases from 2.458 Å to 2.543 Å for the 4 × 4 graphene supercell doped with six B atoms (corresponding to 18.75% B concentration), as observed in other B-doped graphene systems. Similar to graphene systems doped with four B atoms, the considered 4 × 4 graphene supercell doped with six B atoms at adjacent positions experience less structural distortion (Figure 14a) compared to systems doped with three and five B atoms.

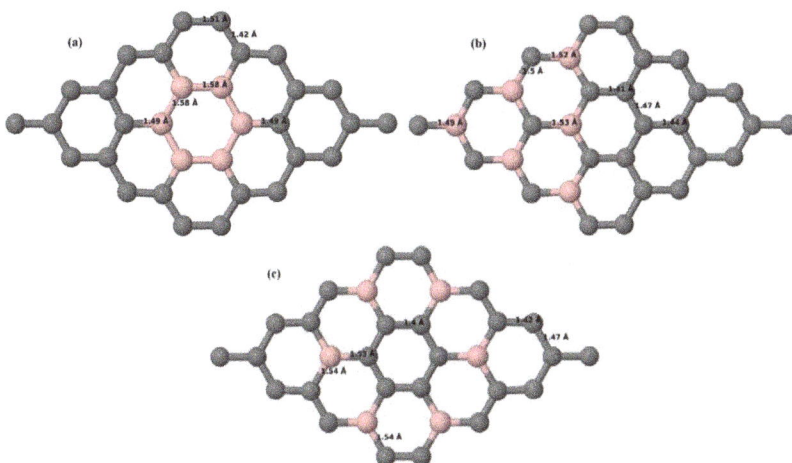

Figure 14. (a–c) Optimized structures of the 4 × 4 graphene supercell doped with six B atoms (18.75% B concentration) with different doping configurations.

The band structures of all graphene systems doped with six B atoms shown in Figure 15a–c indicate *p*-type semiconducting behaviors. At a 18.75% concentration, B-doped graphene does not show a band gap (Figure 15a,c) when the B atoms are located at adjacent and alternate sublattice sites, whereas a band gap of ~0.99 eV opens up in graphene (Figure 15b) when the six B dopants are located at the same sublattice sites of graphene, as summarized in Table 6. The observed closed band gap in graphene doped with six B atoms (Figure 15a,c) upon placing the dopants at adjacent and alternate sublattices could be attributed to the symmetry formed by the B dopants in two triangular sublattices [30].

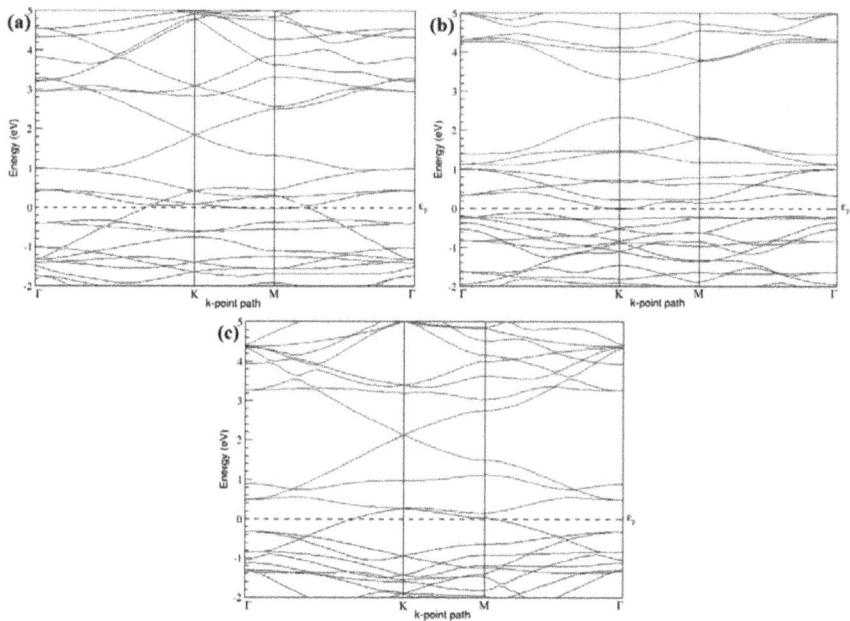

Figure 15. (a–c) Band structures of the 4 × 4 graphene supercell doped with six B atoms (18.75% B concentration) corresponding to the optimized structures shown in Figure 14a–c.

Table 6. The B concentration, doping configurations with considered sublattices, cohesive energies, and the band gaps introduced for 4 × 4 supercell doped with six B atoms.

Model	B Concentration (%)	Configuration	Considered Sublattices for Dopants	E_{coh} (eV/atom)	Band Gap (eV)
		Figure 14a	"B", "A", "B", "A", "B" and "A" (adjacent)	−8.478	0
4 × 4	18.75	Figure 14b	All in "B" (same)	−8.593	0.988
		Figure 14c	"A", "B", "B", "A", "A" and "B" (alternate)	−8.661	0

3.3. N-Doped Graphene

3.3.1. N-Doped Graphene System with One N Atom per Supercell

Here we consider one N-substitutional dopant in 2 × 2, 3 × 3, 4 × 4, and 6 × 6 graphene supercells. Similar to that of graphene systems doped with one B atom, all relaxed graphene systems doped with one N atom remain planar (Figure 16a–d) due to the similar size of the C and introduced N atom, showing good accordance with both theoretical [30,37] and experimental results [53]. Similar to the B atom, as the N atom also undergoes sp^2 hybridization and strongly binds with the three neighboring C atoms through σ-bonds, there is no distortion in the graphene lattice after N doping. The optimized lattice constant decreases from 2.458 Å to 2.456 Å, 2.454 Å, 2.450 Å, and 2.441 Å for 6 × 6, 4 × 4,

3 × 3, and 2 × 2 graphene supercells doped with one N atom (1.39%, 3.13%, 5.56%, and 12.5% N concentrations), respectively. This decrease in the optimized lattice constant with the increase in the N-doping concentration is due to the smaller covalent radius of N compared to C, which is in agreement with previous reports [30].

Figure 16a–d depict the optimized geometries of 2 × 2, 3 × 3, 4 × 4, and 6 × 6 graphene supercells doped with one N atom. The small covalent radius of N compared to that of C results in the reduction of the C–N bond length to 1.41 Å for 2 × 2, 3 × 3, 4 × 4, and 6 × 6 graphene supercells doped with one N atom (Figure 16a–d), consistent with that reported in [37]. The adjoining C–C bond length was reduced from 1.42 Å to 1.41 Å in graphene systems doped with one N atom (Figure 16a–d) to preserve the planar geometry.

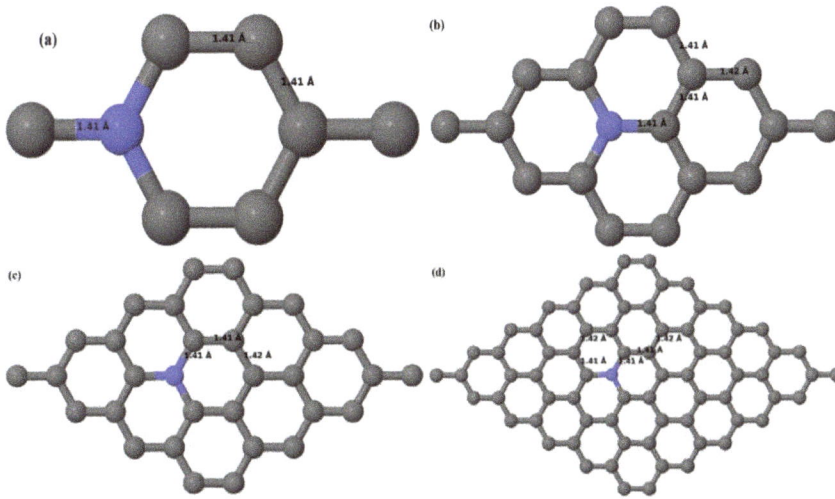

Figure 16. Optimized structures of various graphene systems doped with one N atom, dark blue represents N atoms; (**a**) 2 × 2 graphene supercell with 12.5% N concentration; (**b**) 3 × 3 graphene supercell with 5.56% N concentration; (**c**) 4 × 4 graphene supercell with 3.13% N concentration; (**d**) 6 × 6 graphene supercell with 1.39% N concentration.

Figure 17a–d depict the band structures computed for the relaxed geometries of different graphene systems doped with one N atom shown in Figure 16a–d, in which the linear energy dispersion at the Dirac point is seen unaffected. The observation of the preserved linear energy dispersion near the Dirac point is in good agreement with previous theoretical results [37]. The obtained band structures are compared with those reported in earlier works [30,37,43] and are found to be in excellent agreement.

The 2 × 2 and 4 × 4 graphene supercells doped with one N atom (corresponding to 12.5% and 3.13% N concentrations) show band gaps of ~0.67 eV and ~0.20 eV, respectively, as evident from Figure 17a,c, due to the symmetry breaking of graphene sublattices similar to that observed in corresponding B-doped graphene systems. The band gap value observed for a 2 × 2 graphene supercell doped with one N atom is in agreement with the existing value of 0.67 eV [44]. Similar to 3 × 3 and 6 × 6 supercells doped with one B atom, there is no band gap opening for 3 × 3 and 6 × 6 supercells doped with one N atom (Figure 17b,d). The observed zero band gap in 3 × 3 and 6 × 6 graphene supercells doped with one N atom is consistent with similar theoretical reports [42,44]. For other supercells, the energy gap increases from ~0.20 eV to ~0.67 eV for 4 × 4 and 2 × 2 supercells doped with one N atom (corresponding to N concentrations of 3.13% and 12.5%), respectively (Table 7).

All these graphene systems doped with one N atom exhibit *n*–type metallic nature as evident from the band structures in Figure 17, which is consistent with that reported by Wang et al. [43]. Both 2 × 2 and 4 × 4 graphene supercells doped with one N atom exhibit *n*-type metallic character with band gap values listed in Table 7.

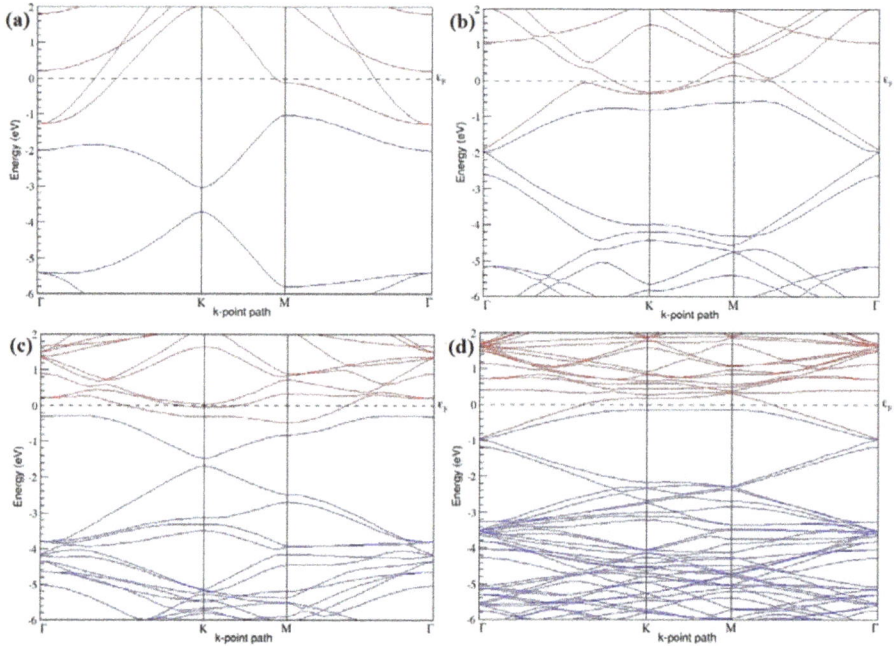

Figure 17. Band structures of graphene systems doped with one N atom corresponding to the optimized structures shown in Figure 16a–d: (**a**) 2 × 2 graphene supercell with 12.5% N concentration; (**b**) 3 × 3 graphene supercell with 5.56% N concentration; (**c**) 4 × 4 graphene supercell with 3.13% N concentration; (**d**) 6 × 6 graphene supercell with 1.39% N concentration.

Table 7. The N concentrations, cohesive energies, and the band gaps introduced for various supercells doped with one N atom.

Model	N Concentration (%)	E_{coh} (eV/atom)	Band Gap (eV)
2 × 2	12.5	−9.029	0.668
3 × 3	5.56	−9.193	0
4 × 4	3.13	−9.254	0.202
6 × 6	1.39	−9.291	0

3.3.2. N-Doped Graphene System with Two N Atoms per Supercell

Here the substitution of two C atoms by two N atoms in 2 × 2, 3 × 3, 4 × 4, and 6 × 6 graphene supercells (Figure 18a–l) are considered. Similar to graphene systems doped with one N atom, the optimized structures of graphene systems doped with two N atoms preserve the planar geometry of PG (Figure 18a–l) even after the introduction of two N atoms. The relaxed lattice constant decreases from 2.458 Å to 2.454 Å, 2.449 Å, 2.441 Å, and 2.422 Å for 6 × 6, 4 × 4, 3 × 3, and 2 × 2 graphene supercells doped with two N atoms (2.78%, 6.25%, 11.11%, 25% N concentrations), respectively, which shows a decrease in lattice constant with increasing N-doping concentration, similar to that observed for graphene systems doped with one N atom.

Figure 18a–l present the optimized geometries of the 2 × 2, 3 × 3, 4 × 4, and 6 × 6 graphene supercells doped with two N atoms, in which the same configurations taken for graphene systems doped with two B atoms are considered. Three doping configurations of 2 × 2, 3 × 3, 4 × 4, and 6 × 6 supercells doped with two N atoms (corresponding to 25%, 11.11%, 6.25%, and 2.78% B concentrations, respectively) are selected, with dopant atoms at adjacent, (Figure 18a,d,g,j), same (Figure 18b,e,h,k), and alternate sublattice points (Figure 18c,f,i,l). In graphene systems doped with two N atoms, the C–N and C–C bond lengths were reduced significantly (Figure 18a–l) as compared to graphene systems doped with one N atom for retaining the structure. After obtaining the stable geometries, the cohesive energies were calculated for all considered graphene systems doped with two N atoms and are listed in Table 8.

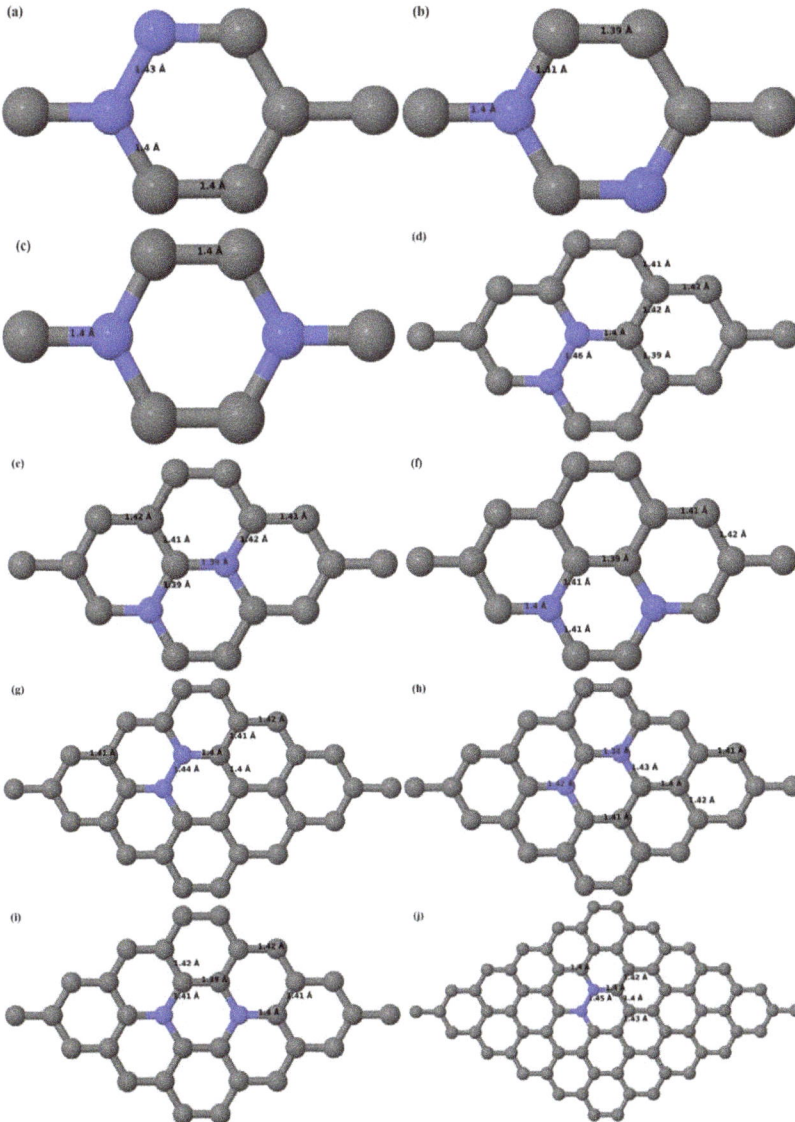

Figure 18. *Cont.*

(k)   (l)

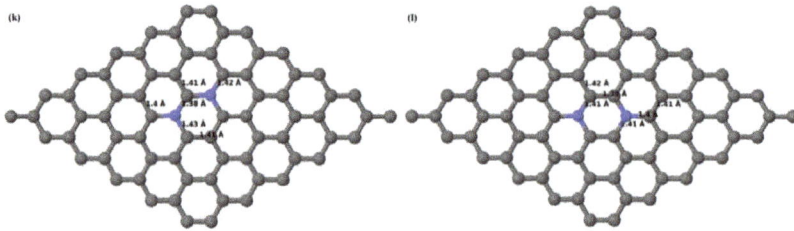

Figure 18. Optimized structures of various graphene systems doped with two N atoms with different doping configurations; (**a–c**) 2 × 2 graphene supercell with 25% N concentration; (**d–f**) 3 × 3 graphene supercell with 11.11% N concentration; (**g–i**) 4 × 4 graphene supercell with 6.25% N concentration; (**j–l**) 6 × 6 graphene supercell with 2.78% N concentration.

Table 8. The N concentrations, doping configurations with considered sublattice, cohesive energies, and the band gaps introduced for various supercells doped with two N atoms.

Model	N Concentration (%)	Configuration	Considered Sublattices for Dopants	E_{coh} (eV/atom)	Band Gap (eV)
2 × 2	25	Figure 18a	"B" and "A" (adjacent)	−8.717	0.400
		Figure 18b	Both in "B" (same)	−8.780	1.324
		Figure 18c	"B" and "A" (alternate)	−8.875	0
3 × 3	11.11	Figure 18d	"B" and "A" (adjacent)	−9.048	0.252
		Figure 18e	Both in "B" (same)	−9.079	0.271
		Figure 18f	"B" and "A" (alternate)	−9.112	0.039
4 × 4	6.25	Figure 18g	"B" and "A" (adjacent)	−9.172	0.174
		Figure 18h	Both in "B" (same)	−9.191	0.403
		Figure 18i	"B" and "A" (alternate)	−9.203	0.012
6 × 6	2.78	Figure 18j	"B" and "A" (adjacent)	−9.253	0.085
		Figure 18k	Both in "B" (same)	−9.262	0.110
		Figure 18l	"B" and "A" (alternate)	−9.268	0.023

Figure 19a–l present the band structures computed for the optimized structures of different graphene systems doped with two N atoms shown in Figure 18a–l. The band structures of 2 × 2, 3 × 3, and 4 × 4 graphene supercells doped with two N atoms are found to be in good agreement with those reported in [43]. Similar to systems doped with two B atoms, the linear dispersion near the Dirac point is not completely destroyed (Figure 19a–l), but an energy band gap opens in all cases except for the 2 × 2 graphene supercell doped with two N atoms at the alternate sublattice points (Figure 19c). At a 25% N concentration, N-doped graphene has band gaps of 0.40 eV (Figure 19a) and ~1.32 eV (Figure 19b) upon placing the N atoms at adjacent and same sublattice positions in graphene, respectively. However, the band gap is found to be closed (Figure 19c) for the configuration in which the N atoms are at alternate sublattices of a 2 × 2 graphene supercell, even though it corresponds to a high N-doping concentration of 25%.

In 4 × 4 graphene supercells doped with two N atoms (corresponding to 6.25% N concentration), band gaps of ~0.17 eV (Figure 19g), ~0.40 eV (Figure 19h), and ~0.01 eV (Figure 19i) are observed for the configurations with N atoms at adjacent, same, and alternate sublattices, respectively. The large band gap opening in 2 × 2 and 4 × 4 graphene supercells doped with two N atoms, with the positioning of the dopant atoms at the same sublattice, is due to the combined effect of the symmetry breaking of sublattices, similar to that observed for graphene systems doped with two B atoms and are in accordance with that reported in [30].

Similar to the 3 × 3 and 6 × 6 graphene supercells doped with two B atoms, the band structures of 3 × 3 and 6 × 6 graphene supercells doped with two N atoms are characterized by non-zero band gaps as shown in Figure 19d–f,j–l. The 3 × 3 graphene supercell doped with two N atoms (corresponding to 11.11% N concentration) exhibit band gaps of ~0.25 eV (Figure 19d), ~0.27 eV (Figure 19e) and ~0.04 eV (Figure 19f) for the configuration with N atoms at adjacent, same, and alternate sublattices, respectively.

In N-doped graphene with 2.78% N concentration, band gaps of 0.085 eV (Figure 19j), ~0.11 eV (Figure 19k), and ~0.02 eV (Figure 19l) appear when the dopants are at adjacent, same, and alternate sublattice sites in graphene.

All these systems doped with two N atoms in 2 × 2, 3 × 3, 4 × 4, and 6 × 6 graphene supercells exhibit *n*-type metallic behavior. Our analysis of the electronic properties of the 2 × 2 and 4 × 4 graphene sheets doped with two N atoms are contrary to that of [43]. Table 8 presents a summary of the considered N-doping concentrations with the doping configurations, the sublattices selected for the N atoms, the calculated cohesive energies, and the band gaps observed in each of these cases.

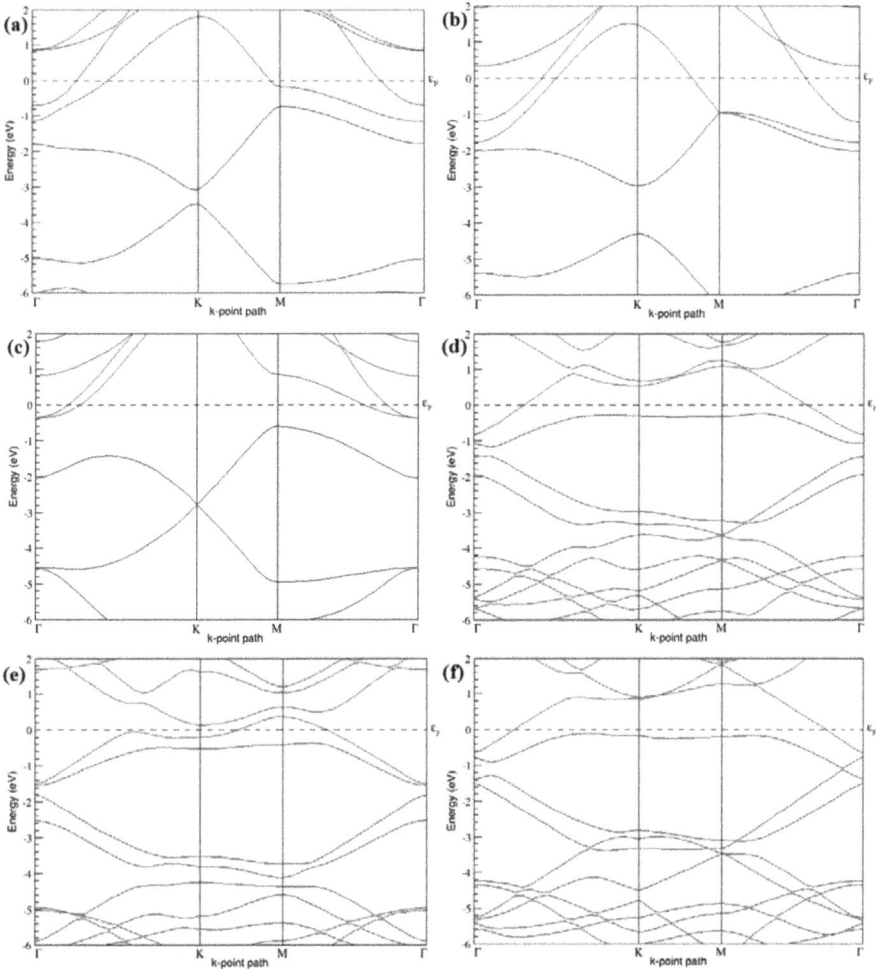

Figure 19. *Cont.*

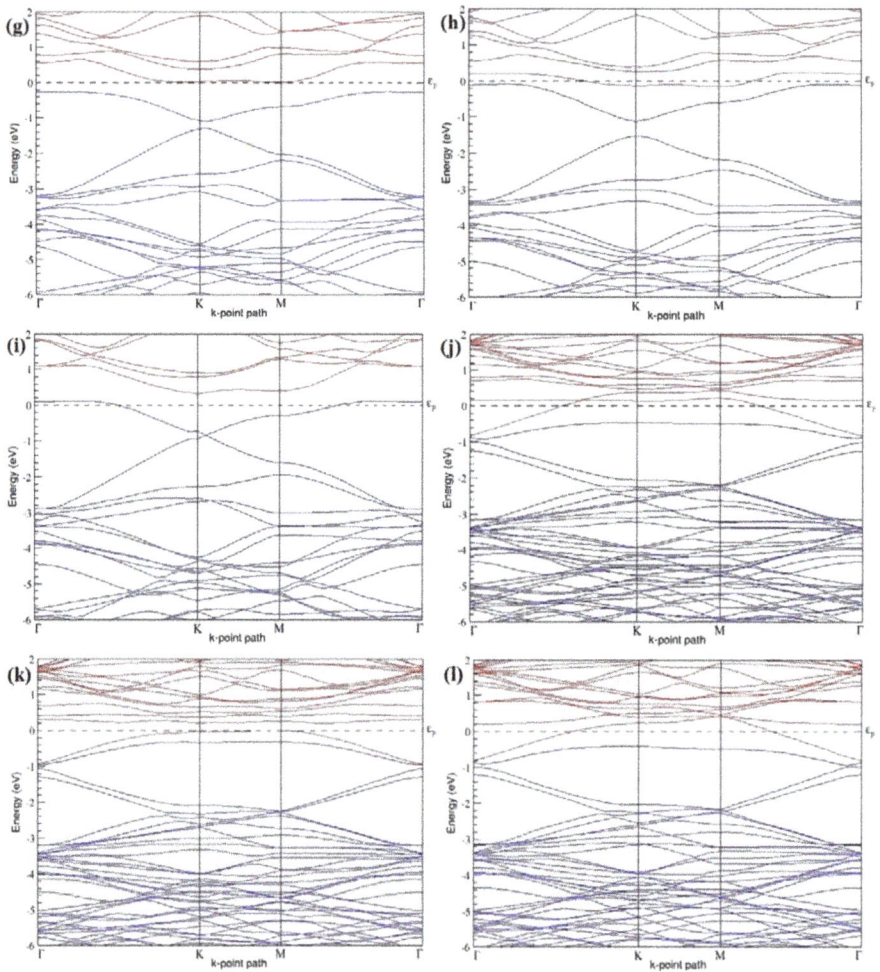

Figure 19. Band structures of graphene systems doped with two N atoms corresponding to the optimized structures shown in Figure 18a–l; (**a–c**) 2 × 2 graphene supercell with 25% N concentration; (**d–f**) 3 × 3 graphene supercell with 11.11% N concentration; (**g–i**) 4 × 4 graphene supercell with 6.25% N concentration; (**j–l**) 6 × 6 graphene supercell with 2.78% N concentration.

3.3.3. N-Doped Graphene System with Three N Atoms per Supercell

Here the substitutions of three C atoms by three N atoms are considered in 3 × 3 and 4 × 4 graphene supercells. All graphene systems doped with three N atoms have planar hexagonal structures as seen in Figure 20a–f. The relaxed lattice constant decreases from 2.458 Å to 2.445 Å and 2.434 Å for 4 × 4 and 3 × 3 graphene supercells doped with three N atoms, (9.38% and 16.67% N concentrations), respectively, similar to that observed in graphene systems doped with one or two N atoms.

Figure 20a–f present the relaxed geometries of 3 × 3 and 4 × 4 graphene supercells doped with three N atoms, with N atoms at adjacent, same, and alternate sublattices, respectively. Compared with their B counterparts, graphene systems doped with three N atoms are found to have significantly less structural distortion, even when the three N atoms are placed at adjacent locations (Figure 20a,d) due to the comparable size of C and N atoms.

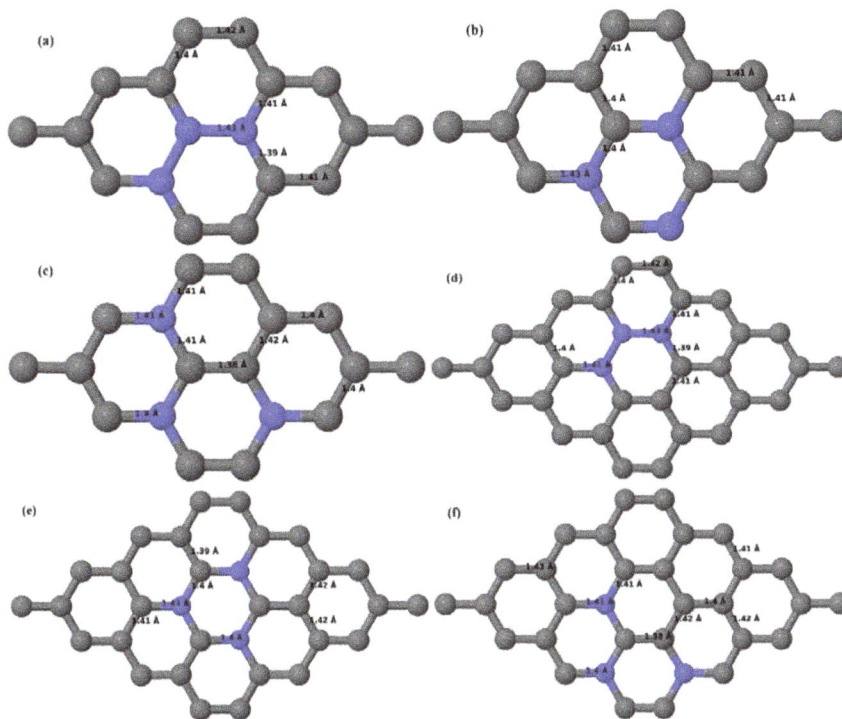

Figure 20. Optimized structures of various graphene systems doped with three N atoms with different doping configurations; (**a–c**) 3 × 3 graphene supercell with 16.67% N concentration; (**d–f**) 4 × 4 graphene supercell with 9.38% N concentration.

Figure 21a–f present the band structures computed for the optimized structures of different graphene systems doped with three N atoms shown in Figure 20a–f. Graphene systems doped with three N atoms in 3 × 3 and 4 × 4 graphene supercells exhibit *n*-type metallic behavior with band gaps indicated in Table 9. Similar to graphene systems doped with three B atoms, with dopants at adjacent positions, deformed band structures were obtained (Figure 21a,d) for those structures in which the N-substitutional dopants are placed at the adjacent positions (Figure 20a,d).

Table 9. The N concentrations, doping configurations with considered sublattice, cohesive energies, and the band gaps introduced for various supercells doped with three N atoms.

Model	N Concentration (%)	Configuration	Considered Sublattices for Dopants	E_{coh} (eV/atom)	Band Gap (eV)
		Figure 20a	"B", "A" and "B" (adjacent)	−8.876	0.294
3 × 3	16.67	Figure 20b	All in "B" (same)	−8.957	0.930
		Figure 20c	"B", "B" and "A" (alternate)	−8.984	0.160
		Figure 20d	"B", "A" and "B" (adjacent)	−9.072	0.270
4 × 4	9.38	Figure 20e	All in "B" (same)	−9.120	0.608
		Figure 20f	"B", "B" and "A" (alternate)	−9.132	0.176

In 3 × 3 graphene systems doped with three N atoms (corresponding to 16.67% N concentration), a band gap of 0.93 eV (Figure 21b) appears when the dopant atoms are placed at the same sublattices, whereas band gap values of ~0.29 eV (Figure 21a) and 0.16 eV (Figure 21c) appear when the dopant atoms are placed at the adjacent and alternate sublattice positions, respectively. The 4 × 4 graphene supercell doped with three N atoms (corresponding to 9.38% N concentration) induce band gaps of 0.27 eV (Figure 21d), ~0.61 eV (Figure 21e), and ~0.18 eV (Figure 21f),for the configuration with N atoms at adjacent, same, and alternate sublattices, respectively.

Table 9 presents the doping concentrations, considered doping concentrations with sublattices, cohesive energies, and the band gaps introduced for all graphene systems doped with three N atoms.

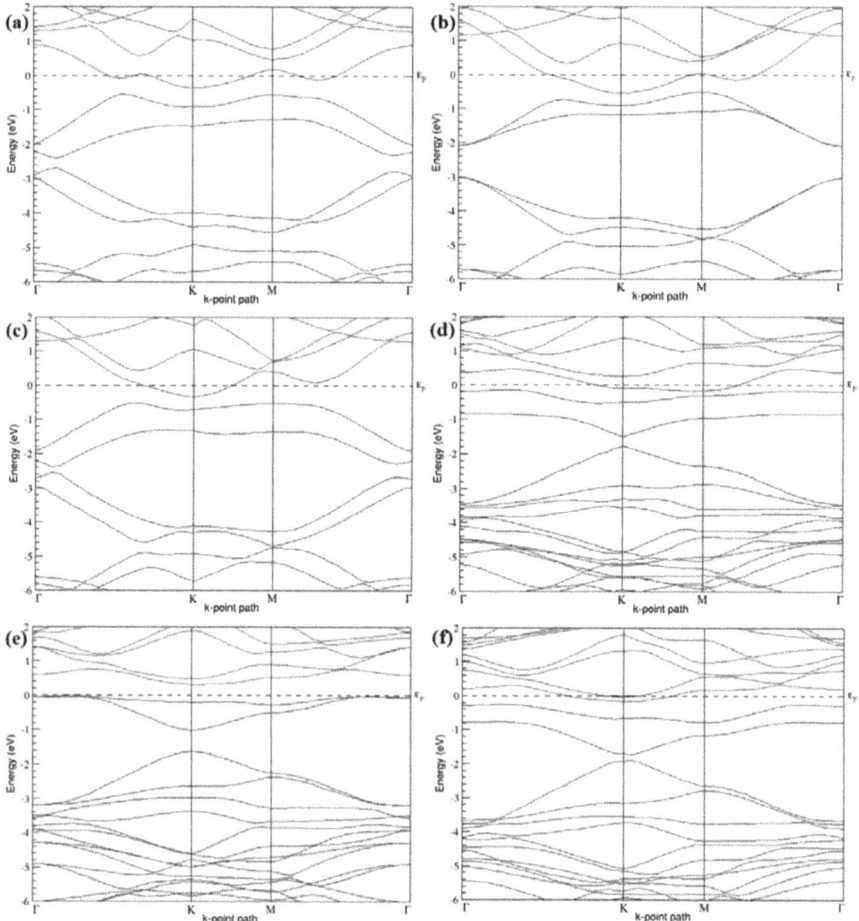

Figure 21. Band structures of graphene systems doped with three N atoms corresponding to the configurations shown in Figure 20a–f; (a–c) 3 × 3 graphene supercell with 16.67% N concentration; (d–f) 4 × 4 graphene supercell with 9.38% N concentration.

3.3.4. N-Doped Graphene System with Four N Atoms per Supercell

Here we consider the substitution of four C atoms by four N atoms in 3 × 3 and 4 × 4 graphene supercells. After structural relaxation, all graphene systems doped with four N atoms appear to be planar by adjusting the adjoining bond lengths (Figure 22a–f). The relaxed lattice constant decreases from 2.458 Å to 2.441 Å and 2.427 Å for 4 × 4 and 3 × 3 graphene supercells doped with four N atoms (corresponding to 12.5% and 22.22% N concentrations), respectively, which also indicates a decrease in lattice constant with the increase in N-doping concentration, as observed with other B-doped graphene systems.

Figure 22a–f present the relaxed structures of 3 × 3 and 4 × 4 supercells doped with four N atoms at adjacent, same, and alternate sublattices, respectively. As compared to graphene systems doped with four B atoms, graphene systems doped with four N atoms experience almost negligible structural distortion even upon placing the N atoms at the adjacent positions in the lattice (Figure 22a,d).

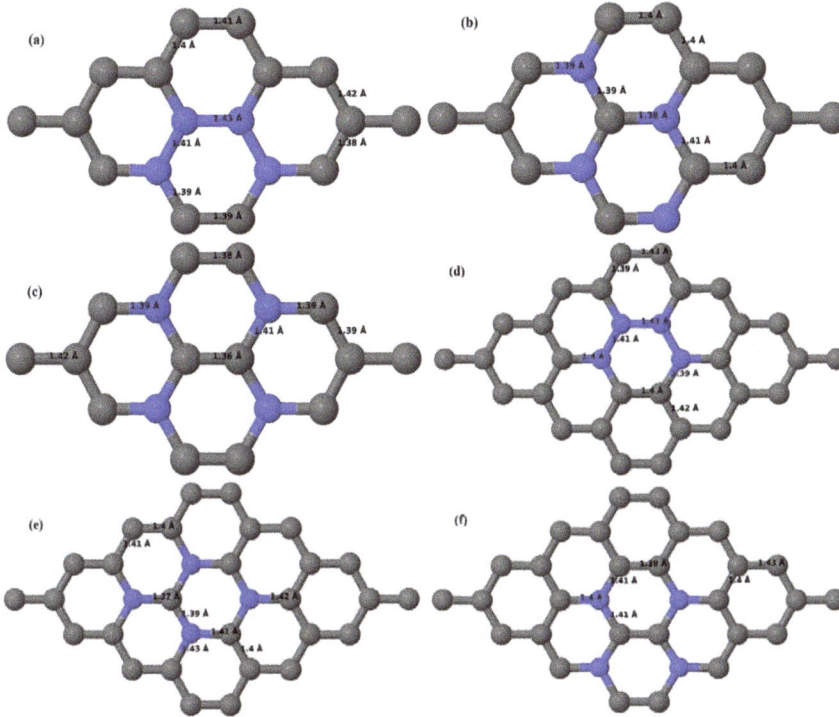

Figure 22. Optimized structures of various graphene systems doped with four N atoms with different doping configurations; (a–c) 3 × 3 graphene supercell with 22.22% N concentration; (d–f) 4 × 4 graphene supercell with 12.5% N concentration.

Figure 23a–f present the band structures computed for the optimized structures of different graphene systems doped with four N atoms shown in Figure 22a–f. Similar to graphene systems doped with three N atoms, all four N-doped graphene structures exhibit *n*-type metallic properties with band gaps as seen in Figure 23a–f. In 3 × 3 graphene systems doped with four N atoms (corresponding to 22.22%), the highest band gap value of ~0.80 eV (Figure 23b) appears when the dopant atoms are placed at the same sublattices, whereas band gaps of 0.20 eV (Figure 23a) and ~0.11 eV (Figure 23c) are induced in graphene when the dopant atoms are placed at the adjacent positions and alternate sublattices, respectively. At a 12.5% N concentration, a maximum band gap of 0.70 eV (Figure 23e) opens up when the dopant atoms are at the same sublattice, and a minimum band gap value of ~0.004 eV (Figure 23f) appears when the dopant atoms are at alternate sublattice positions. The observed very small band gap for the doping configuration with N atoms at adjacent locations (Figure 23d) corresponding to 12.5% N-doping concentration could be ascribed to the symmetry formed by the N dopants in the two triangular sublattices.

Table 10 presents the doping concentrations with selected sublattices, the cohesive energies, and the band gaps introduced for all graphene systems doped with four N atoms.

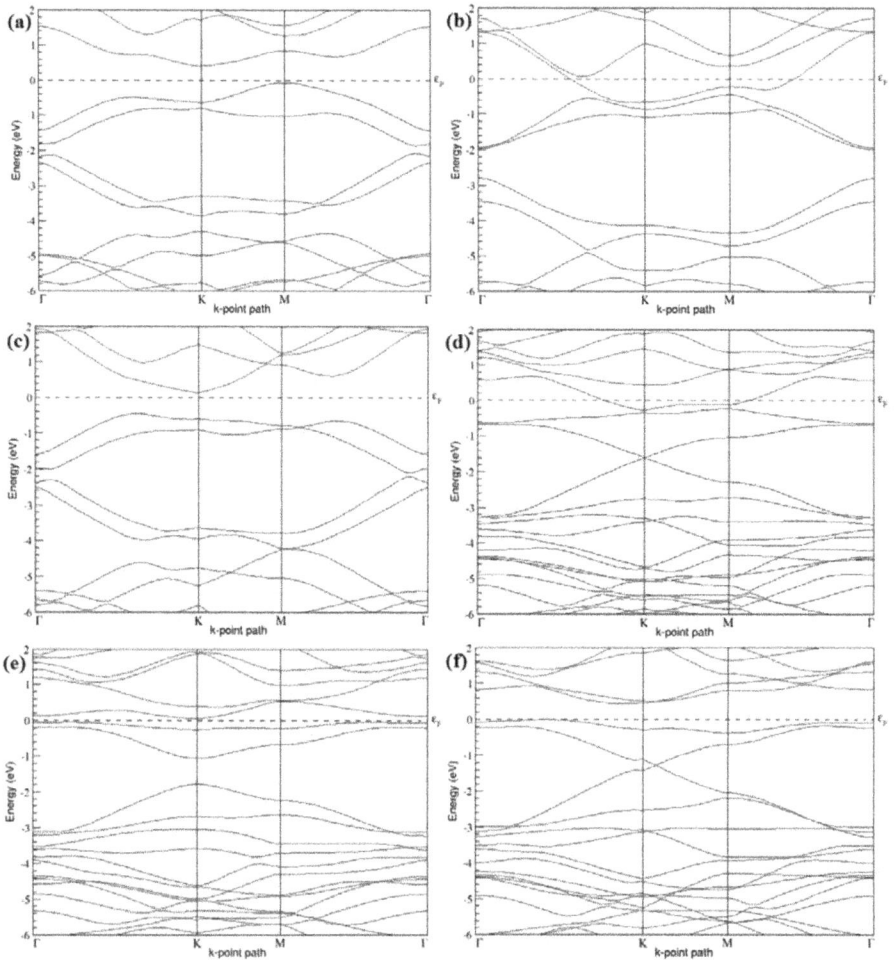

Figure 23. Band structures of graphene systems doped with four N atoms corresponding to the optimized structures shown in Figure 22a–f; (**a–c**) 3 × 3 graphene supercell with 22.22% N concentration; (**d–f**) 4 × 4 graphene supercell with 12.5% N concentration.

Table 10. The N concentrations, doping configurations with considered sublattice, cohesive energies, and the band gaps introduced for various supercells doped with four N atoms.

Model	N Concentration (%)	Configuration	Considered Sublattices for Dopants	E_{coh} (eV/atom)	Band Gap (eV)
		Figure 22a	"B", "A", "B" and "A" (adjacent)	−8.741	0.200
3 × 3	22.22	Figure 22b	All in "B" (same)	−8.844	0.804
		Figure 22c	"B", "B", "A" and "A" (alternate)	−8.894	0.106
		Figure 22d	"B", "A", "B" and "A" (adjacent)	−8.987	0.020
4 × 4	12.5	Figure 22e	All in "A" (same)	−9.052	0.700
		Figure 22f	"B", "B", "A" and "A" (alternate)	−9.077	0.004

3.3.5. N-Doped Graphene System with Five N Atoms per Supercell

Here the substitution of five C atoms by five N atoms in a 4 × 4 graphene supercell are considered, with the dopant atoms at adjacent, same, and alternate sublattice positions. After structural relaxation, it was observed that all graphene systems doped with five N atoms exhibit planar geometry (Figure 24a–c), similar to other N-doped graphene systems. The relaxed lattice constant decreases from 2.458 Å to 2.435 Å for the 4 × 4 graphene supercell doped with five N atoms (corresponding to 15.63% N concentration), as observed in other N-doped graphene systems. Similar to graphene systems doped with three N atoms, there is no structural distortion in the considered structures of graphene systems doped with five N atoms. The planar configuration is maintained by adjusting the associated bond lengths.

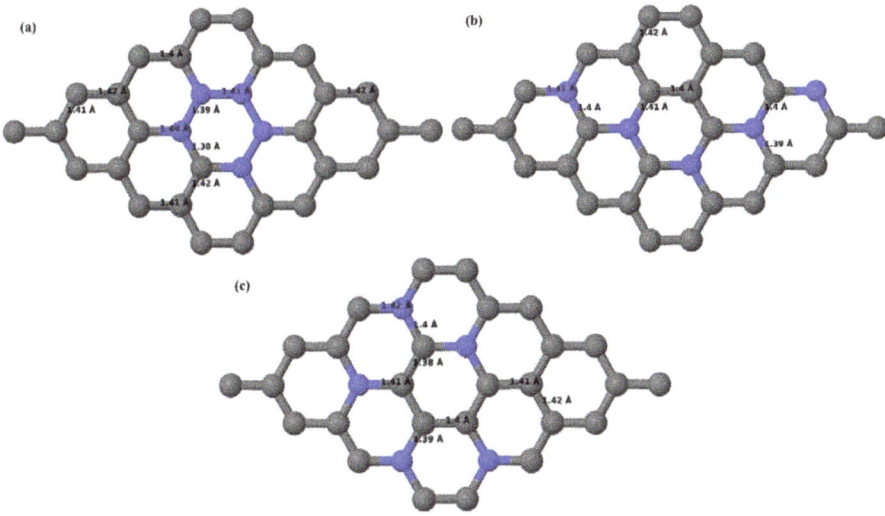

Figure 24. (a–c) Optimized structures of the 4 × 4 graphene supercell doped with five N atoms (15.63% N concentration) with different doping configurations.

The band structures presented in Figure 25a–c indicate that all graphene structures doped with five N atoms show an *n*-type metallic character. At a 15.63% N concentration, N atoms located at adjacent, same, and alternate sublattices induce band gaps of ~0.36 eV (Figure 25a), ~0.79 eV (Figure 25b), and ~0.14 eV (Figure 25c), respectively, in graphene as summarized in Table 11. Table 11 summarizes the observed band gaps and the calculated cohesive energies for different doping configurations corresponding to the 15.63% N concentration in graphene.

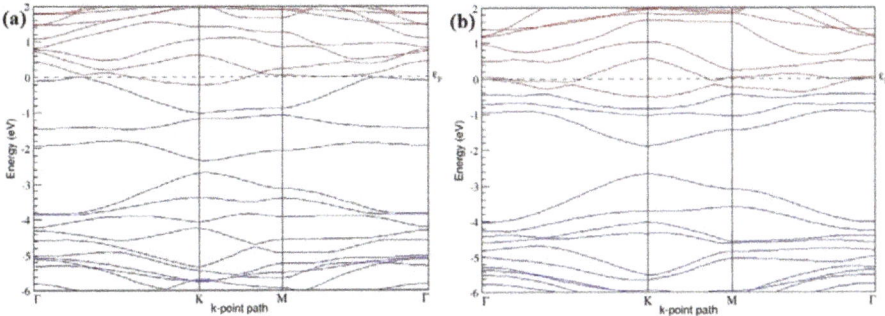

Figure 25. *Cont.*

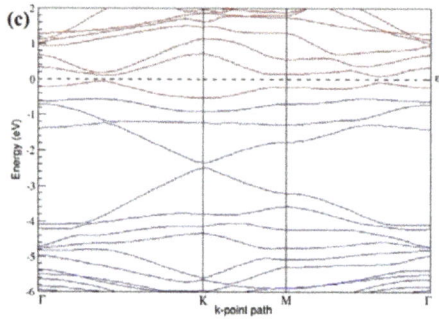

Figure 25. (a–c) Band structures of 4 × 4 graphene supercell doped with five N atoms (15.63% N concentration) corresponding to the optimized structures shown in Figure 24a–c.

Table 11. The N concentrations, doping configurations with considered sublattice, cohesive energies, and the band gaps introduced for 4 × 4 supercell doped with five N atoms.

Model	N Concentration (%)	Configuration	Considered Sublattices for Dopants	E_{coh} (eV/atom)	Band Gap (eV)
		Figure 24a	"B", "A", "B", "A" and "B" (adjacent)	−8.884	0.357
4 × 4	15.63	Figure 24b	All in "B" (same)	−8.988	0.792
		Figure 24c	"B", "A", "B", "A" and "B" (alternate)	−9.017	0.141

3.3.6. N-Doped Graphene System with Six N Atoms per Supercell

Here the substitutions of six C atoms by six N atoms are considered in the 4 × 4 graphene supercell with the dopant atoms at adjacent, same, and alternate sublattice positions. The planar structure is preserved even after the introduction of six N atoms in the lattice (Figure 26a–c). The relaxed lattice constant decreases from 2.458 Å to 2.431 Å for 4 × 4 graphene supercell doped with six N atoms (corresponding to 18.75% N concentration), as observed in other N-doped graphene systems. Similar to other N-doped graphene systems, there is no geometrical distortion in the graphene systems doped with six N atoms.

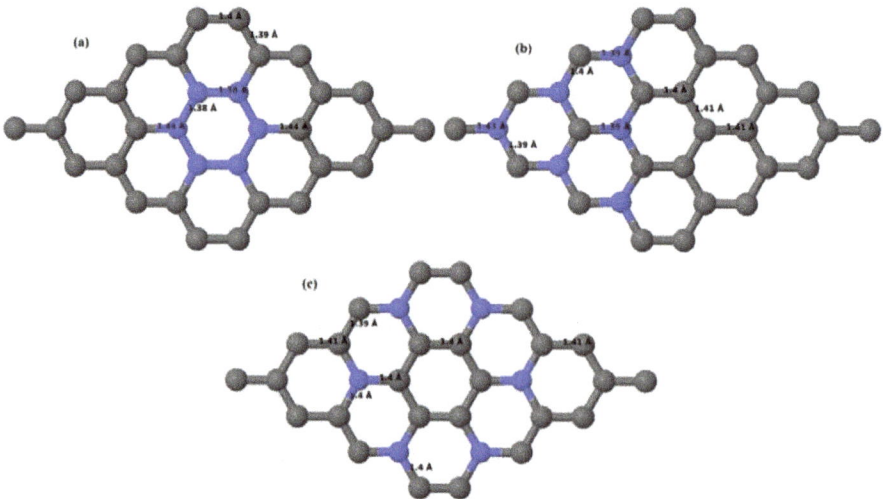

Figure 26. (a–c) Optimized structures of the 4 × 4 graphene supercell doped with six nitrogen atoms (18.75% N concentration) with different doping configurations.

All these 4 × 4 graphene systems doped with six N atoms show *n*-type metallic behavior. At an 18.75% N concentration, a band gap of ~1.04 eV (Figure 27b) opens up in graphene, when the six N atoms are located at the same sublattice sites of graphene. Upon placing the N atoms at adjacent and alternate sublattices, there is no band gap opening as evident from Figure 27a,c, similar to that observed in graphene systems doped with six B atoms. Table 12 summarizes the observed band gaps and the calculated cohesive energies for different doping configurations corresponding to the 18.75% N concentration in graphene.

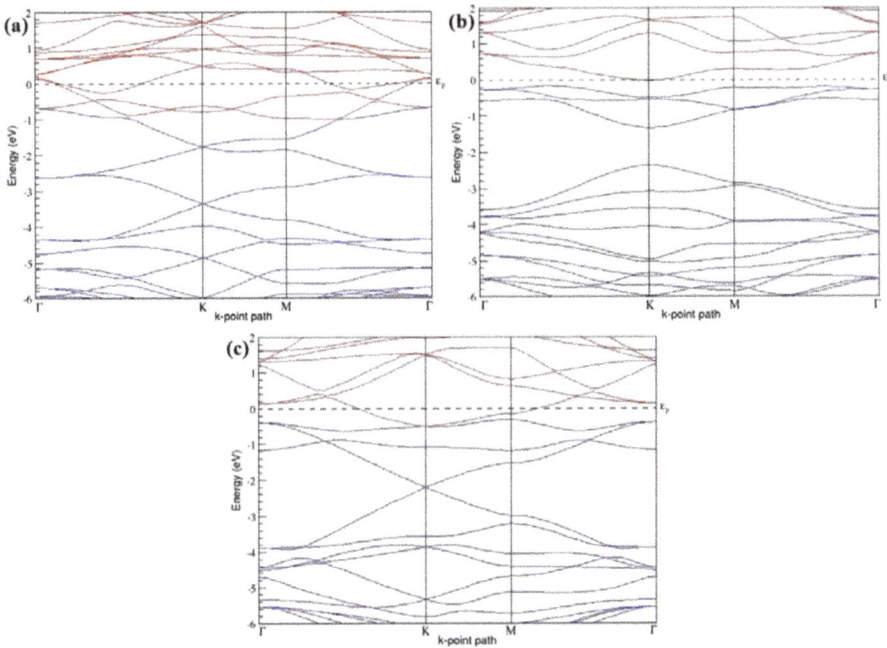

Figure 27. (a–c) Band structures of the 4 × 4 graphene supercell doped with six N atoms (18.75% N concentration) corresponding to the optimized structures shown in Figure 26a–c.

Table 12. The N concentrations, doping configurations with considered sublattice, cohesive energies, and the band gaps introduced for 4 × 4 supercell doped with six N atoms.

Model	N Concentration (%)	Configuration	Considered Sublattices for Dopants	E_{coh} (eV/atom)	Band Gap (eV)
		Figure 26a	"B", "A", "B", "A", "B" and "A" (adjacent)	−8.753	0
4 × 4	18.75	Figure 26b	All in "B" (same)	−8.920	1.042
		Figure 26c	"A", "B", "B", "A", "A" and "B" (alternate)	−8.976	0

4. Discussions

The changes in the bond lengths after B-doping can be read from the optimized structures shown in Figures 3, 6, 8, 10, 12, and 14. In all B-doped graphene systems, the bond lengths are adjusted to retain the planar geometry, i.e., with long C–B bonds and relatively short adjacent C–C bonds based on the number and location of B atoms in graphene supercells. The placement of B atoms at adjacent locations results in large B–B bond lengths in B-doped graphene systems. The large B–B bond lengths observed for the configurations with B atoms at adjacent locations (Figures 6a,d,g,j, 8a,d, 10a,d, 12a, and 14a could be ascribed to the large covalent radius of B compared to the host C atoms. The changes in the bond lengths after N-doping can be read from the optimized structures shown in Figures 16, 18,

20, 22, 24, and 26. The planar configuration is also maintained in all N-doped graphene systems by adjusting the associated bond lengths, i.e., with short C–N and C–C bond lengths.

The obtained negative values of cohesive energies presented in Tables 1–12 indicate that all considered B- and N-doped graphene systems are energetically stable. It was also observed that the cohesive energy increases with increase in B- and N-doping concentration (Tables 1–12), which shows that energetic stability decreases with increasing B- and N-doping concentration. For the same doping concentration, the cohesive energy was found to be lowest for the doping configuration with dopants at an alternate sublattice and highest for the doping configuration with dopants at adjacent positions. These results show that B- and N-doped graphene with dopants placed at alternate sublattices are more stable than that with dopants placed at adjacent and same sublattice positions.

Tables 1–12 show that the band gap in general increases with the increase in the B- and N-doping concentration for the doping configuration with the highest band gap, which is in agreement with previous reports [30]. Except in 3 × 3 and 6 × 6 graphene supercells doped with two B atoms, the band gap was found to be at a maximum when the B atoms were at the same sublattice and at a minimum when they were at adjacent or alternate sublattice positions, which is in agreement with report by Rani et al. [30]. Even though 3 × 3 and 6 × 6 graphene supercells doped with two B atoms show a band gap between the valence and conduction bands, no specific dependence of the band gap on the B-atom positioning could be highlighted. Apart from the presented isomers of 3 × 3 and 6 × 6 graphene supercells doped with two B atoms (corresponding to 11.11% and 2.78% B concentrations), the band structures of several other isomers for the doping configurations of adjacent, same, and alternate sublattices are also calculated (which are not presented here) to check for the observed variation of band gap dependency on dopant positioning from that found in other B-doped graphene systems. The calculated band gaps in all cases of 3 × 3 and 6 × 6 graphene supercells doped with two B atoms do not follow any specific dependency on the B-doping configurations. The nature of this anomaly is still being investigated. On the other hand, 3 × 3 graphene supercells doped with three and four B or N atoms showed the highest band gap for the doping configurations with dopant atoms at same sublattices and the lowest band gap for the doping configuration with dopant atoms at alternate sublattices or adjacent positions. The trend of the band gap dependency on the doping configurations in all N-doped graphene systems was observed to be similar to that observed in corresponding B-doped graphene systems. Similar to the exception observed in the case of the 3 × 3 and 6 × 6 graphene supercells doped with two B atoms, no specific correlation between the band gap and the doping configuration could be determined for 3 × 3 and 6 × 6 graphene supercells doped with two N atoms (corresponding to 11.11% and 2.78% N concentrations). The band gaps of 2 × 2 graphene supercell doped with two N atoms (1.323 eV at a 25% N concentration) and 4 × 4 graphene supercell doped with six N atoms (1.042 eV at an 18.75% N concentration) are found to be very close to the band gap of silicon (1.1 eV).

The observed band gaps for the considered B or N concentrations corresponding to the most stable configurations are summarized in Table 13. For the most probable configuration corresponding to the doping concentrations ranging from 2.78% to 22.22%, the observed band gaps lie within the range of 0.008–0.190 eV and 0.004–0.202 eV in B- and N-doped systems, respectively, while the most stable configuration corresponding to B or N concentrations of 1.39%, 5.56%, 18.75%, and 25% do not show a band gap. Thus, the band gap of graphene could be tailored from 0.008 eV to 0.190 eV by B-doping and from 0.004 eV to 0.202 eV by N-doping, within the concentration range of 2.78%–22.22%, excluding 5.56% and 18.75% concentrations.

Table 13. The doping concentrations, the most stable doping configuration, and the observed band gaps.

Concentration (%)	Most Stable Configuration (B-Doped)	Band Gap (eV) (B-Doped)	Most Stable Configuration (N-Doped)	Band Gap (eV) (N-Doped)
1.39	Figure 3d	0	Figure 16d	0
2.78	Figure 6l	0.008	Figure 18l	0.023
3.13	Figure 3c	0.190	Figure 16c	0.202
5.56	Figure 3b	0	Figure 16b	0
6.25	Figure 6i	0.047	Figure 18i	0.012
9.38	Figure 8f	0.156	Figure 20f	0.176
11.11	Figure 6f	0.046	Figure 18f	0.039
12.5	Figure 10f	0.050	Figure 22f	0.004
15.63	Figure 12c	0.133	Figure 24c	0.141
16.67	Figure 8c	0.107	Figure 20c	0.160
18.75	Figure 14c	0	Figure 26c	0
22.22	Figure 10c	0.062	Figure 22c	0.106
25	Figure 6c	0	Figure 18c	0

As 2×2 graphene supercells doped with one B atom or one N atom and 4×4 graphene supercells doped with four B atoms or four N atoms correspond to the same doping concentration of 12.5%, the cohesive energies and electronic band structures of three different configurations of the 4×4 graphene sheet doped with four B atoms or four N atoms were compared with those of the 2×2 graphene sheet doped with one B atom or one N atom. The widths of the band gaps for a 12.5% doping concentration simulated by a 2×2 graphene supercell doped with one B atom or one N atom and a 4×4 graphene supercell doped with four B atoms or four N atoms were found to be almost the same only for the configuration of a 4×4 graphene supercell doped with four B atoms or four N atoms at the same sublattice. The band gaps introduced for other configurations of the 4×4 graphene supercell doped with four B atoms or four N atoms (adjacent and alternate sublattice positions) were observed to be comparatively small. However, the comparison of the cohesive energies of the 2×2 graphene sheet doped with one B atom or one N atom (-8.795 eV/atom and -9.029 eV/atom, respectively) and the 4×4 graphene supercell doped with four B atoms or four N atoms at the same sublattice (-8.829 eV/atom and -9.052 eV/atom, respectively) indicate that the cohesive energy is strongly dependent on the supercell size considered for the doping concentration. Thus, to achieve a 12.5% B or N doping concentration, the 4×4 graphene supercell doped with four B atoms or four N atoms is more stable than the 2×2 graphene supercell doped with one B atom or one N atom.

5. Conclusions

We have here calculated the energetic stabilities and the structural and electronic properties of B- and N-doped graphene with varying doping concentrations and several doping configurations in different graphene supercell sizes, using first-principles density functional theory calculations. It was observed that both B- and N-doped graphene maintain the planar geometry of pristine graphene with a slight distortion with longer C–B bonds and shorter C–N bonds compared to the C–C bond length, which is in agreement with previous reports. The doped structures with dopant atoms placed at adjacent locations have been found to be highly distorted, with less distortion in N-doped graphene systems compared to that of the corresponding B-doped graphene systems.

The stability was found to be decreasing with increases in B- and N-doping concentrations. For a particular doping concentration, stability was found to be higher for the atomic configuration with dopant atoms at alternate sublattice positions than other configurations and decreases in the order of alternate > same > adjacent. The cohesive energies of the N-doped graphene systems were found to be lower than that of similar B-doped graphene systems; hence, N-doped graphene structures are considered to be more stable than their B-doped counterparts. As the doping concentration decreases, the cohesive energy difference between similar N- and B-doped graphene structures also decrease, which indicates that graphene structures with a light doping of B and N atoms are highly stable.

All B-doped graphene systems exhibit *p*-type doping electronic properties as the Fermi level moves into the valence band, whereas all N-doped graphene systems exhibit *n*-type doping electronic properties as the Fermi level moves into the conduction band. The results also show that graphene with one B- and N-atom doping in any size of supercell exhibits *p*-type semiconducting and *n*-type metallic characters, respectively, with or without band gaps based on the doping concentration. The analysis of the electronic structures of 3×3 and 6×6 graphene supercells with more than one B or N atoms have shown non-zero band gaps around the Dirac point, different from the zero band gap observed for 3×3 and 6×6 graphene supercells doped with one B atom or one N atom. It was also observed that, for the same B- and N-doping concentration, the distribution of the dopant atoms in the crystal lattice determines the width of the introduced band gap around the Dirac point and affects the electronic band structures. The calculations show that B- and N-doped graphene systems with more than one dopant atom exhibit *p*-type semiconducting and *n*-type metallic behavior, respectively, with or without band gaps based on the doping concentrations and doping configurations.

Except in 3×3 and 6×6 graphene supercells with two B and N dopants (corresponding to 11.11% and 2.87% B and N doping concentrations, respectively), the band gap dependency on the dopant sites was observed to be the same, where a maximum band gap opens up when the dopant atoms are at the same sublattice and a minimum band gap opens up when the dopants are at alternate sublattices. No conceivable correlation between the band gap and the doping configurations could be deduced from the band structures of graphene doped with two B atoms and two N atoms in 3×3 and 6×6 graphene supercells. The 3×3 graphene supercells doped with three B atoms, three N atoms, four B atoms, or four N atoms show similar band gap dependency on the dopant locations that has been observed with other supercells which are not multiples of three.

Our calculations indicate that band gap can be adjusted as required based on the doping concentration and the doping configuration, which is of great significance in designing graphene-based semiconductor devices.

Supplementary Materials: The following are available online at www.mdpi.com/2079-9292/5/4/91/s1, Figure S1: Convergence curve showing total energy against the plane wave cutoff energy; Figure S2: Convergence curve showing total energy against k-point grid.

Author Contributions: Seba Sara Varghese performed the ABINIT simulations, analyzed the results, drawn conclusions and wrote the paper. Sundaram Swaminathan, Krishna Kumar Singh and Vikas Mittal supervised the work. All authors discussed the results and commented on the manuscript at all stages.

Conflicts of Interest: The authors declare no conflict of interest.

References

1. Allen, M.J.; Tung, V.C.; Kaner, R.B. Honeycomb carbon: A review of graphene. *Chem. Rev.* **2010**, *110*, 132–145. [CrossRef] [PubMed]
2. Novoselov, K.S.; Geim, A.K.; Morozov, S.V.; Jiang, D.; Zhang, Y.; Dubonos, S.V.; Grigorieva, I.V.; Firsov, A.A. Electric field effect in atomically thin carbon films. *Science* **2004**, *306*, 666–669. [CrossRef] [PubMed]
3. Castro Neto, A.H.; Guinea, F.; Peres, N.M.R.; Novoselov, K.S.; Geim, A.K. The electronic properties of graphene. *Rev. Mod. Phys.* **2009**, *81*, 109–162. [CrossRef]
4. Du, X.; Skachko, I.; Barker, A.; Andrei, E.Y. Approaching ballistic transport in suspended graphene. *Nat. Nanotechnol.* **2008**, *3*, 491–495. [CrossRef] [PubMed]
5. Bolotin, K.I.; Sikes, K.J.; Jiang, Z.; Klima, M.; Fudenberg, G.; Hone, J.; Kim, P.; Stormer, H.L. Ultrahigh electron mobility in suspended graphene. *Solid State Commun.* **2008**, *146*, 351–355. [CrossRef]
6. Bolotin, K.I.; Ghahari, F.; Shulman, M.D.; Stormer, H.L.; Kim, P. Observation of the fractional quantum hall effect in graphene. *Nature* **2011**, *475*, 122. [CrossRef]
7. Geim, A.K.; Novoselov, K.S. The rise of graphene. *Nat. Mater.* **2007**, *6*, 183–191. [CrossRef] [PubMed]
8. Avouris, P.; Chen, Z.; Perebeinos, V. Carbon-based electronics. *Nat. Nanotechnol.* **2007**, *2*, 605–615. [CrossRef] [PubMed]
9. Geim, A.K. Graphene: Status and prospects. *Science* **2009**, *324*, 1530–1534. [CrossRef] [PubMed]
10. Wallace, P.R. The band theory of graphite. *Phys. Rev.* **1947**, *71*, 622–634. [CrossRef]

11. Charlier, J.C.; Michenaud, J.P.; Gonze, X.; Vigneron, J.P. Tight-binding model for the electronic properties of simple hexagonal graphite. *Phys. Rev. B* **1991**, *44*, 13237–13249. [CrossRef]

12. Mak, K.F.; Lui, C.H.; Shan, J.; Heinz, T.F. Observation of an electric-field-induced band gap in bilayer graphene by infrared spectroscopy. *Phys. Rev. Lett.* **2009**, *102*, 256405. [CrossRef] [PubMed]

13. Boukhvalov, D.W.; Katsnelson, M.I. Tuning the gap in bilayer graphene using chemical functionalization: Density functional calculations. *Phys. Rev. B* **2008**, *78*, 085413. [CrossRef]

14. Han, M.Y.; Özyilmaz, B.; Zhang, Y.; Kim, P. Energy band-gap engineering of graphene nanoribbons. *Phys. Rev. Lett.* **2007**, *98*, 206805. [CrossRef] [PubMed]

15. Denis, P.A. Band gap opening of monolayer and bilayer graphene doped with aluminium, silicon, phosphorus, and sulfur. *Chem. Phys. Lett.* **2010**, *492*, 251–257. [CrossRef]

16. Dai, J.Y.; Yuan, J.M.; Giannozzi, P. Gas adsorption on graphene doped with B, N, Al, and S: A theoretical study. *Appl. Phys. Lett.* **2009**, *95*, 232105. [CrossRef]

17. Zhang, Y.H.; Chen, Y.B.; Zhou, K.G.; Liu, C.H.; Zeng, J.; Zhang, H.L.; Peng, Y. Improving gas sensing properties of graphene by introducing dopants and defects: A first-principles study. *Nanotechnology* **2009**, *20*, 185504. [CrossRef] [PubMed]

18. Lv, R.; Li, Q.; Botello-Mendez, A.R.; Hayashi, T.; Wang, B.; Berkdemir, A.; Hao, Q.Z.; Elias, A.L.; Cruz-Silva, R.; Gutierrez, H.R.; et al. Nitrogen-doped graphene: Beyond single substitution and enhanced molecular sensing. *Sci. Rep.* **2012**, *2*, 586. [CrossRef] [PubMed]

19. Niu, F.; Tao, L.-M.; Deng, Y.-C.; Wang, Q.-H.; Song, W.-G. Phosphorus doped graphene nanosheets for room temperature nh3 sensing. *New J. Chem.* **2014**, *38*, 2269–2272. [CrossRef]

20. Kwon, O.S.; Park, S.J.; Hong, J.Y.; Han, A.R.; Lee, J.S.; Lee, J.S.; Oh, J.H.; Jang, J. Flexible fet-type vegf aptasensor based on nitrogen-doped graphene converted from conducting polymer. *ACS Nano* **2012**, *6*, 1486–1493. [CrossRef] [PubMed]

21. Niu, F.; Liu, J.-M.; Tao, L.-M.; Wang, W.; Song, W.-G. Nitrogen and silica co-doped graphene nanosheets for NO_2 gas sensing. *J. Mater. Chem. A* **2013**, *1*, 6130–6133. [CrossRef]

22. Yang, G.-H.; Zhou, Y.-H.; Wu, J.-J.; Cao, J.-T.; Li, L.-L.; Liu, H.-Y.; Zhu, J.-J. Microwave-assisted synthesis of nitrogen and boron co-doped graphene and its application for enhanced electrochemical detection of hydrogen peroxide. *RSC Adv.* **2013**, *3*, 22597–22604. [CrossRef]

23. Wu, Z.S.; Ren, W.C.; Xu, L.; Li, F.; Cheng, H.M. Doped graphene sheets as anode materials with superhigh rate and large capacity for lithium ion batteries. *ACS Nano* **2011**, *5*, 5463–5471. [CrossRef] [PubMed]

24. Feng, Y.Q.; Tang, F.L.; Lang, J.W.; Liu, W.W.; Yan, X.B. Facile approach to preparation of nitrogen-doped graphene and its supercapacitive performance. *J. Inorg. Mater.* **2013**, *28*, 677–682. [CrossRef]

25. Ma, C.C.; Shao, X.H.; Cao, D.P. Nitrogen-doped graphene nanosheets as anode materials for lithium ion batteries: A first-principles study. *J. Mater. Chem.* **2012**, *22*, 8911–8915. [CrossRef]

26. Fang, H.; Yu, C.; Ma, T.; Qiu, J. Boron-doped graphene as a high-efficiency counter electrode for dye-sensitized solar cells. *Chem. Commun.* **2014**, *50*, 3328–3330. [CrossRef] [PubMed]

27. Gadipelli, S.; Guo, Z.X. Graphene-based materials: Synthesis and gas sorption, storage and separation. *Progress Mater. Sci.* **2015**, *69*, 1–60. [CrossRef]

28. Kemp, K.C.; Chandra, V.; Saleh, M.; Kim, K.S. Reversible CO_2 adsorption by an activated nitrogen doped graphene/polyaniline material. *Nanotechnology* **2013**, *24*, 235703. [CrossRef] [PubMed]

29. Zhou, Y.G.; Zu, X.T.; Gao, F.; Nie, J.L.; Xiao, H.Y. Adsorption of hydrogen on boron-doped graphene: A first-principles prediction. *J. Appl. Phys.* **2009**, *105*, 014309. [CrossRef]

30. Rani, P.; Jindal, V.K. Designing band gap of graphene by b and n dopant atoms. *RSC Adv.* **2013**, *3*, 802–812. [CrossRef]

31. Lazar, P.; Zboril, R.; Pumera, M.; Otyepka, M. Chemical nature of boron and nitrogen dopant atoms in graphene strongly influences its electronic properties. *Phys. Chem. Chem. Phys.* **2014**, *16*, 14231–14235. [CrossRef] [PubMed]

32. Zhang, W.; Wu, L.; Li, Z.; Liu, Y. Doped graphene: Synthesis, properties and bioanalysis. *RSC Adv.* **2015**, *5*, 49521–49533. [CrossRef]

33. Huang, B. Electronic properties of boron and nitrogen doped graphene nanoribbons and its application for graphene electronics. *Phys. Lett. A* **2011**, *375*, 845–848. [CrossRef]

34. Martins, T.B.; Miwa, R.H.; da Silva, A.J.R.; Fazzio, A. Electronic and transport properties of boron-doped graphene nanoribbons. *Phys. Rev. Lett.* **2007**, *98*. [CrossRef] [PubMed]

35. Lherbier, A.; Blase, X.; Niquet, Y.-M.; Triozon, F.; Roche, S. Charge transport in chemically doped 2D graphene. *Phys. Rev. Lett.* **2008**, *101*, 036808. [CrossRef] [PubMed]

36. Panchokarla, L.S.; Subrahmanyam, K.S.; Saha, S.K.; Govindaraj, A.; Krishnamurthy, H.R.; Waghmare, U.V.; Rao, C.N.R. Synthesis, structure, and properties of boron- and nitrogen-doped graphene. *Adv. Mater.* **2009**, *21*, 4726–4730. [CrossRef]

37. Wu, M.; Cao, C.; Jiang, J.Z. Light non-metallic atom (B, N, O and F)-doped graphene: A first-principles study. *Nanotechnology* **2010**, *21*, 505202. [CrossRef] [PubMed]

38. Garcia-Lastra, J.M. Strong dependence of band-gap opening at the dirac point of graphene upon hydrogen adsorption periodicity. *Phys. Rev. B* **2010**, *82*, 235418. [CrossRef]

39. Cruz-Silva, E.; Barnett, Z.M.; Sumpter, B.G.; Meunier, V. Structural, magnetic, and transport properties of substitutionally doped graphene nanoribbons from first principles. *Phys. Rev. B* **2011**, *83*, 155445. [CrossRef]

40. Mukherjee, S.; Kaloni, T.P. Electronic properties of boron- and nitrogen-doped graphene: A first principles study. *J. Nanopart. Res.* **2012**, *14*, 1059. [CrossRef]

41. Wang, Z.Y.; Xiao, J.R.; Li, X.Y. Effects of heteroatom (boron or nitrogen) substitutional doping on the electronic properties of graphene nanoribbons. *Solid State Commun.* **2012**, *152*, 64–67. [CrossRef]

42. Woinska, M.; Milowska, K.; Majewski, J.A. Ab initio modeling of graphene functionalized with boron and nitrogen. *Acta Phys. Pol. A* **2012**, *122*, 1087–1089. [CrossRef]

43. Wang, Z.; Qin, S.; Wang, C. Electronic and magnetic properties of single-layer graphene doped by nitrogen atoms. *Eur. Phys. J. B* **2014**, *87*, 1–6. [CrossRef]

44. Ye-Cheng, Z.; Hao-Li, Z.; Wei-Qiao, D. A 3N rule for the electronic properties of doped graphene. *Nanotechnology* **2013**, *24*, 225705.

45. Panchakarla, L.S.; Govindaraj, A.; Rao, C.N.R. Boron- and nitrogen-doped carbon nanotubes and graphene. *Inorg. Chim. Acta* **2010**, *363*, 4163–4174. [CrossRef]

46. Fujimoto, Y. Formation, energetics, and electronic properties of graphene monolayer and bilayer doped with heteroatoms. *Adv. Condens. Matter Phys.* **2015**, *2015*, 571490. [CrossRef]

47. Usachov, D.; Vilkov, O.; Grüneis, A.; Haberer, D.; Fedorov, A.; Adamchuk, V.K.; Preobrajenski, A.B.; Dudin, P.; Barinov, A.; Oehzelt, M.; et al. Nitrogen-doped graphene: Efficient growth, structure, and electronic properties. *Nano Lett.* **2011**, *11*, 5401–5407. [CrossRef] [PubMed]

48. Gonze, X.; Amadon, B.; Anglade, P.M.; Beuken, J.M.; Bottin, F.; Boulanger, P.; Bruneval, F.; Caliste, D.; Caracas, R.; Cote, M.; et al. Abinit: First-principles approach to material and nanosystem properties. *Comput. Phys. Commun.* **2009**, *180*, 2582–2615. [CrossRef]

49. Perdew, J.P.; Burke, K.; Ernzerhof, M. Generalized gradient approximation made simple. *Phys. Rev. Lett.* **1996**, *77*, 3865–3868. [CrossRef] [PubMed]

50. Troullier, N.; Martins, J.L. Efficient pseudopotentials for plane-wave calculations. *Phys. Rev. B* **1991**, *43*, 1993–2005. [CrossRef]

51. Monkhorst, H.J.; Pack, J.D. Special points for brillouin-zone integrations. *Phys. Rev. B* **1976**, *13*, 5188–5192. [CrossRef]

52. Dresselhaus, M.S.; Dresselhaus, G.; Saito, R.; Jorio, A. Raman spectroscopy of carbon nanotubes. *Phys. Rep.* **2005**, *409*, 47–99. [CrossRef]

53. Wang, Y.; Shao, Y.; Matson, D.W.; Li, J.; Lin, Y. Nitrogen-doped graphene and its application in electrochemical biosensing. *ACS Nano* **2010**, *4*, 1790–1798. [CrossRef] [PubMed]

electronics

MDPI

Article

Guided Modes in a Double-Well Asymmetric Potential of a Graphene Waveguide

Yi Xu and Lay Kee Ang *

Engineering Product Development, Singapore University of Technology and Design, Singapore 487372, Singapore; yi_xu@mymail.sutd.edu.sg
* Correspondence: ricky_ang@sutd.edu.sg; Tel.: +65-6499-4558

Academic Editors: Yoke Khin Yap and Zhixian Zhou
Received: 27 October 2016; Accepted: 1 December 2016; Published: 7 December 2016

Abstract: The analogy between the electron wave nature in graphene electronics and the electromagnetic waves in dielectrics has suggested a series of optical-like phenomena, which is of great importance for graphene-based electronic devices. In this paper, we propose an asymmetric double-well potential on graphene as an electronic waveguide to confine the graphene electrons. The guided modes in this graphene waveguide are investigated using a modified transfer matrix method. It is found that there are two types of guided modes. The first kind is confined in one well, which is similar to the asymmetric quantum well graphene waveguide. The second kind can appear in two potential wells with double-degeneracy. Characteristics of all the possible guide modes are presented.

Keywords: guided modes; graphene; electronic waveguide

---

1. Introduction

A two-dimensional layer of carbon atoms known as graphene [1] has been investigated widely both theoretically and experimentally. Graphene has a unique band structure for which the electron and hole bands meet at two inequivalent points in the Brillouin zone. At these Dirac points, the electrons behave like quasi-particles according to the massless Dirac equation, which leads to a linear dispersion relation. The analogy between the electron wave nature in graphene electronics and the electromagnetic waves in dielectrics has suggested a series of optical-like phenomena, such as the Goos–Hänchen effect [2,3], negative refraction [4], collimation [5], birefringence [6], and the Bragg reflection [7] reported in recent papers.

Another analogy is the graphene-based electron waveguides [8–18], which will be useful for various graphene-based devices, such as electronic fiber [19]. The crux of such an electronic waveguide is the confinement of Dirac fermions in graphene. There are many schemes to confine electrons in graphene, e.g., electric confinement [20,21], magnetic confinement [22] and strain-induced confinement [16–18]. By having a quantum well in graphene to confine massless Dirac fermions, the guided modes in graphene-based waveguides with quantum well structure induced by an external electrical field have been investigated in detail [8–10]. Experimentally, gate-controlled electron guiding in graphene by tuning the carrier type and its density using local electrostatic fields has been achieved [12]. However, the confinement of the quasi-particles is not strong enough due to the Klein tunneling.

Another effective way to confine the electrons in graphene is by applying a uniform magnetic field [22] and the characteristics of magnetic waveguides in graphene have also been studied [13,14]. Recently, the strain-induced graphene waveguides has also attracted much attention [16–18]. Contrary to the electric or magnetic waveguide in graphene, the strain-induced waveguide confines electrons without any external fields. It confines the quasi-particles with the pseudo-magnetic field [23], which arises from the applied mechanical strain. The bound states of the strain-induced waveguides are dependent on the valleys, which is different from the electric and magnetic waveguides (valley independent). Furthermore, smooth one-dimensional potential [15,24–26] and velocity barriers [27] have been proposed to confine the Dirac fermions in graphene. Electrons can also be guided in nano-structured graphene, such as graphene nano-ribbons [28] and antidot lattice [29]. Recently, Allen et al. demonstrated the confinement of electron waves in graphene with the use of superconducting interferometry in a graphene Josephson junction [30].

For a given graphene waveguide with quantum well structures, the dispersion equation for its guided modes is normally determined by applying the continuity of wave function at the interfaces of the quantum well. This method is easy for one quantum well structure, but not for multiple or more complicated quantum well structures, e.g., a double-well potential. In this paper, we will apply a transfer matrix method [31–33] to deduce the dispersion equation for guided modes in a double-well potential structure. From the calculation, we will present the characteristics of the guided modes in detail, and report its novel properties as compared to other graphene-based waveguides.

In the present work, we focus on the properties of guided modes in a double-well asymmetric potential that acts as a slab waveguide for electron waves (based on Dirac solution) in a form similar to that in integrated optics. However, it should be noted that the optical properties or optical waves in graphene have also been investigated [34–41]. This work shows an analogy between the electron wave nature in graphene electronics and the electromagnetic waves in dielectrics.

2. Guided Mode and Dispersion Equation for a Double-Well Potential

In the presence of an electrostatic potential $V(x)$, electrons inside a monolayer graphene can be described by the Dirac-like equation:

$$[-i\hbar v_F \vec{\sigma} \cdot \nabla + V(x)]\Psi(x,y) = E\Psi(x,y),$$ (1)

where $\vec{\sigma} = (\sigma_x, \sigma_y)$ are the Pauli matrices, and $v_F = 10^6$ m/s is the Fermi velocity. $\Psi = (\tilde{\Psi}_A, \tilde{\Psi}_B)^T$ is a two-component pseudo-spinor wave function, and $\tilde{\Psi}_{A,B}$ are the smooth enveloping functions for two triangular sublattices in graphene, which can be expressed as $\tilde{\Psi}_{A,B}(x,y) = \Psi_{A,B}(x)e^{ik_y y}$ due to its translation invariance in the direction. In terms of $\Psi_{A,B}$, Equation (1) is written as

$$\frac{d\Psi_{A,B}}{dx} \mp k_y \Psi_{A,B} = i\frac{E - V(x)}{\hbar v_F}\Psi_{B,A}.$$ (2)

By using the transfer matrix method [31–33], we obtain a matrix M to connect the wave functions $\Psi_{A,B}(x)$ at the two boundaries at $x = 0$ and $x = d$:

$$\begin{pmatrix} \Psi_A(d) \\ \Psi_B(d) \end{pmatrix} = M \begin{pmatrix} \Psi_A(0) \\ \Psi_B(0) \end{pmatrix},$$ (3)

$$M = \begin{pmatrix} \dfrac{\cos(k_x d - \theta)}{\cos\theta} & s\dfrac{i\sin(k_x d)}{\cos\theta} \\ s\dfrac{i\sin(k_x d)}{\cos\theta} & \dfrac{\cos(k_x d + \theta)}{\cos\theta} \end{pmatrix},$$ (4)

where $s = \text{sgn}(E - V)$, $k = |E - V|/\hbar v_F$, $(k_x, k_y) = k(\cos\theta, \sin\theta)$. By solving Equation (3), one can immediately obtain the dispersion equation of a graphene waveguide (see below).

Here, we consider a double-well asymmetric potential for graphene waveguide as plotted in Figure 1a, where its potential distribution $V(x)$ is denoted by

$$V(x) = \begin{cases} V_1 & x < 0 \\ V_2 & 0 < x < h_1 \\ V_3 & h_1 < x < (h_1 + h_2) \\ V_4 & x > (h_1 + h_2) \end{cases}. \tag{5}$$

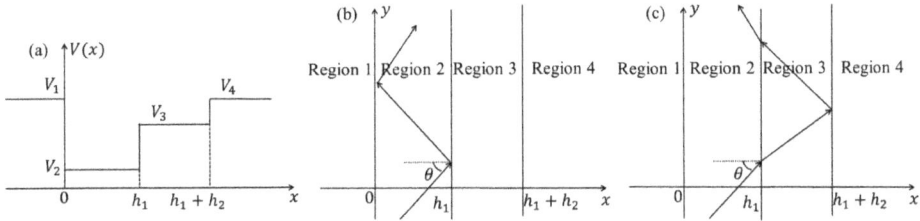

Figure 1. Schematic diagram of (**a**) a double-well potential on graphene; (**b**) guided modes in the region $(0, h_1)$; (**c**) guided modes in the region $(0, h_1 + h_2)$.

The proposed waveguide configuration can be realized by applying four gate-tunable potential barriers on single-layer graphene [42]. By tuning the applied voltage on the gate, a double-well asymmetric potential can be formed on graphene. For simplicity, we have $V_1 = V_4$ unless it is specified elsewhere. In this model, we have neglected the microscopic details of the interaction effects such as the inter-valley coupling [43,44] and inter-valley scattering in the potential steps [44,45]. The electron wave incident into the quantum well (region 2) with an angle θ respect to the x-axis, and the guided modes are propagating in the y-direction. The critical angle of the incident electron waves from region i to region j is defined by

$$\theta_{ij} = \arcsin(|E - V_j|/|E - V_i|), \tag{6}$$

where $i, j = 1, 2, 3, 4(i \neq j)$, E is the energy of incident electron and $V_{i(j)}$ is the electrostatic potential in the respective region $i(j)$. The critical angle θ_{ij} is shown in Figure 2 as a function of the incidence energy E for $V_1 = V_4 = 100$ meV, $V_2 = 0$ meV, $V_3 = 60$ meV, $h_1 = 100$ nm, and $h_2 = 80$ nm.

For a given guided mode, there will be total internal reflection of electron waves at both the two interfaces of a waveguide. For example, at large incidence angles $\theta > \max(\theta_{21}, \theta_{23})$, total reflection occurs in a specific range (marked with "slanted lines") as plotted in Figure 2. In this range, the electron waves will be reflected at the interfaces back and forth with an angle of in region 2 ($0 < x < h_1$) with a guided wave propagating in the axis as shown in Figure 1b.

For the angle within $\theta_{23} > \theta > \theta_{21}$ in region 2 and the angle $\theta_3 > \theta_{34}$ in region 3, which lies in a range marked with "vertical lines" in Figure 2, the electron waves will refract into region 3, and total internal reflection occurs at the interface $x = h_1 + h_2$. The electron wave will oscillate in the region of $0 < x < h_1 + h_2$ as shown in Figure 1c. It is important to note that there are no guided modes within $h_1 < x < h_1 + h_2$, as there is no angle θ to fulfill the condition $\theta_3 > \max(\theta_{32}, \theta_{34})$. Thus, there are only two types of guided modes in a double-well potential. The first one is the guided modes in $0 < x < h_1$, and the other is the guided modes in $0 < x < h_1 + h_2$.

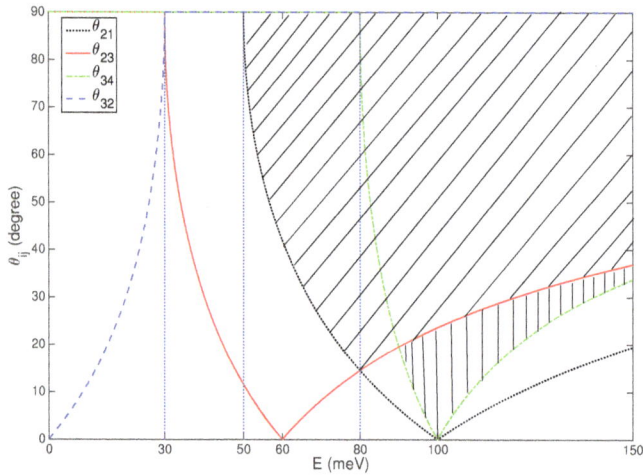

Figure 2. The critical angle θ_{ij} as a function of the incidence energy E. The physical parameters are: $V_1 = V_4 = 100$ meV, $V_2 = 0$ meV, $V_3 = 60$ meV, $h_1 = 100$ nm, $h_2 = 80$ nm.

To derive the dispersion equation, the electron wave function $\Psi_{AB}(x)$ in the double-well potential can be written as:

$$\Psi_A(x) = \begin{cases} A_1 e^{\alpha_1 x} & x < 0 \\ A_2 e^{ik_{2x}(x-h_1)} + B_2 e^{-ik_{2x}(x-h_1)} & 0 < x < h_1 \\ A_3 e^{ik_{3x}(x-h_1-h_2)} + B_3 e^{-ik_{3x}(x-h_1-h_2)} & h_1 < x < (h_1 + h_2) \\ A_4 e^{-\alpha_4(x-h_1-h_2)} & x > (h_1 + h_2) \end{cases} \quad (7)$$

$$\Psi_B(x) = \begin{cases} c_1 A_1 e^{\alpha_1 x} & x < 0 \\ c_2 A_2 e^{ik_{2x}(x-h_1)} - d_2 B_2 e^{-ik_{2x}(x-h_1)} & 0 < x < h_1 \\ c_3 A_3 e^{ik_{3x}(x-h_1-h_2)} - d_3 B_3 e^{-ik_{3x}(x-h_1-h_2)} & h_1 < x < (h_1 + h_2) \\ c_4 A_4 e^{-\alpha_4(x-h_1-h_2)} & x > (h_1 + h_2) \end{cases} \quad (8)$$

Here, we have $k_i = |E - V_i|/\hbar v_F$, $s_i = \text{sgn}(E - V_i)(i = 1, 2, 3, 4)$, $k_y = k_2 \sin \theta$, $k_{2x} = \sqrt{k_2^2 - k_y^2}$, $k_{3x} = \sqrt{k_3^2 - k_y^2}$, $\alpha_1 = \alpha_4 = \sqrt{k_y^2 - k_1^2}$. And $c_1 = -is_1(\alpha_1 - k_y)/k_1$, $c_2 = -is_2(ik_{2x} - k_y)/k_2$, $c_3 = -is_3(ik_{3x} - k_y)/k_3$, $c_4 = is_4(\alpha_4 + k_y)/k_4$, $d_2 = -is_2(ik_{2x} + k_y)/k_2$, $d_3 = -is_3(ik_{3x} + k_y)/k_3$. Based on the transfer matrix method, we have:

$$\begin{pmatrix} 1 \\ c_4 \end{pmatrix} A_4 = \begin{pmatrix} \Psi_A(h_1 + h_2) \\ \Psi_B(h_1 + h_2) \end{pmatrix} = M_3 M_2 \begin{pmatrix} \Psi_A(0) \\ \Psi_B(0) \end{pmatrix} = M_3 M_2 \begin{pmatrix} 1 \\ c_1 \end{pmatrix} A_1, \quad (9)$$

where $M_{2(3)} = \begin{pmatrix} \dfrac{\cos\left(k_{2x(3x)}h_{1(2)} - \theta_{2(3)}\right)}{\cos\theta_{2(3)}} & s_{2(3)}\dfrac{i\sin\left(k_{2x(3x)}h_{1(2)}\right)}{\cos\theta_{2(3)}} \\ \\ s_{2(3)}\dfrac{i\sin\left(k_{2x(3x)}h_{1(2)}\right)}{\cos\theta_{2(3)}} & \dfrac{\cos\left(k_{2x(3x)}h_{1(2)} + \theta_{2(3)}\right)}{\cos\theta_{2(3)}} \end{pmatrix}$, and $\sin\theta_{2(3)} = k_y/k_{2(3)}$.

By multiplying Equation (9) with matrix $(-c_4, 1)$, we obtain

$$\begin{pmatrix} -c_4 & 1 \end{pmatrix} M_3 M_2 \begin{pmatrix} 1 \\ c_1 \end{pmatrix} = 0. \tag{10}$$

From Equation (10), we obtain the dispersion equation for the guided modes in a double-well potential, which is

$$\tan(k_{2x}h_1) = \frac{(k_{2x}h_1)\tan(k_{3x}h_2)f_2 - s_1 s_3 \tan(k_{3x}h_2)f_3 + (k_{2x}h_1)(k_{3x}h_1)f_1}{s_1 s_2 f_4 - (k_{3x}h_2)f_2 - \tan(k_{3x}h_2)f_1 f_2 + s_2 s_3 (k_{2x}h_1)(k_{3x}h_1)\tan(k_{3x}h_2)f_1} = F(k_{2x}h_1), \tag{11}$$

and $f_1 = \sqrt{(k_2 h_1)^2 - (k_{2x}h_1)^2 - (k_1 h_1)^2}$, $f_2 = (k_2 h_1)^2 - (k_{2x}h_1)^2$, $f_3 = (k_1 h_1)(k_{2x}h_1)(k_{3x}h_1)$, $f_4 = (k_1 h_1)(k_2 h_1)(k_{3x}h_1)$, and $k_{3x}h_2 = \sqrt{(k_3 h_1)^2 - (k_2 h_1)^2 + (k_{2x}h_1)^2}h_2/h_1$. The dispersion equation for the guided modes obtained with the transfer matrix method is more convenient than the usual method that solving the Dirac equations from the continuity of wave function at the interfaces of quantum wells. The dispersion equation (Equation (11)) obtained for the double-well asymmetric potential can be recovered to that of the symmetric quantum well-based graphene waveguide by setting $V_3 = V_4 = V_1$ and $h_2 = 0$ nm. Thus, we have

$$\tan(k_{2x}h_1) = \frac{s_1\sqrt{(k_2 h_1)^2 - (k_{2x}h_1)^2 - (k_1 h_1)^2}(k_{2x}h_1)}{s_2(k_1 h_1)(k_2 h_1) - s_1\left((k_2 h_1)^2 - (k_{2x}h_1)^2\right)} = F(k_{2x}h_1), \tag{12}$$

which is the dispersion equation for a symmetric quantum well, as reported in Ref. [8]. Equation (12) can also be directly obtained by using the transfer matrix method:

$$\begin{pmatrix} \Psi_A(h_1) \\ \Psi_B(h_1) \end{pmatrix} = M_2 \begin{pmatrix} \Psi_A(0) \\ \Psi_B(0) \end{pmatrix}. \tag{13}$$

3. Characteristics of the Guided Modes in a Double-Well Potential

For the guided mode in the region $0 < x < h_1$, the electron waves are evanescent in the other three regions, namely $k_y^2 > k_3^2$, where $k_{3x} = \sqrt{k_3^2 - k_y^2}$ is an imaginary number, and k_{2x}, $\alpha_1 = \alpha_4$ are real. Thus, we have the following conditions:

$$\begin{cases} k_{2x} < \sqrt{k_2^2 - k_1^2} \\ k_{2x} < \sqrt{k_2^2 - k_3^2} \end{cases}, \tag{14}$$

which is used to determine the range of k_{2x}.

There are three energy ranges for a guided mode in this region: (i) $E < V_3$; (ii) $V_3 < E < V_1$; and (iii) $E > V_1$, as shown Figure 2 (marked with oblique lines). To show the values of $k_{2x}h_1$, Equation (11) (LHS: $\tan(k_{2x}h_1)$ and RHS: $F(k_{2x}h_1)$) are plotted in Figure 3 as a function of $k_{2x}h_1$ for three energy levels: $E = 56$, 94 and 110 meV. The intersections of $\tan(k_{2x}h_1)$ and $F(k_{2x}h_1)$ give the specific values for the guided modes.

For the $E < V_3$ case, we have $s_1 = s_3 = s_4 = -1$ and $s_2 = 1$, which corresponds to the regime of Klein tunneling. Here, we set $E = 56$ meV, and there is only one intersection ($k_{2x}h_1 = 2.9257$), as shown in Figure 3a. The corresponding wave function distribution is plotted in Figure 4a. Similar to the optical waveguide, the guided modes is defined by the number of the nodes of the spinor wave function $(\Psi_A, -i\Psi_B)^T$. The spinor components Ψ_A and $-i\Psi_B$ represent electron and hole states, respectively. It is clearly seen from Figure 4a that there is no fundamental mode in this case since only Klein tunneling occurs in the double-well potential. The probability current density of the guided mode is plotted in Figure 4b, which can be calculated by the definition in the Dirac equation, $J_y = v_F \Psi^+ \sigma_y \Psi$

with $\sigma_y = \begin{pmatrix} 0 & -i \\ i & 0 \end{pmatrix}$, $\Psi = \begin{pmatrix} \Psi_A(x) \\ \Psi_B(x) \end{pmatrix} e^{ik_y y}$. This finding indicates that the Dirac fermions can be well-localized in the double-well potential.

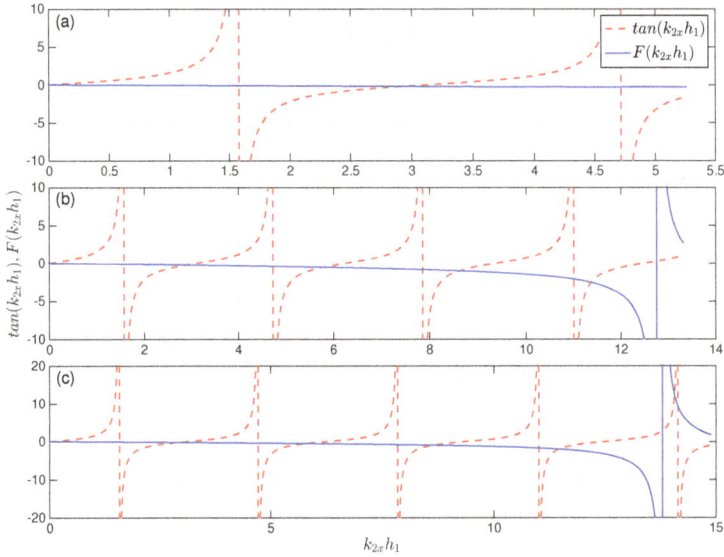

Figure 3. Graphical determination of $k_{2x}h_1$ for guided modes in $(0, h_1)$. The **red** and **blue** curves correspond to $\tan(k_{2x}h_1)$ and $F(k_{2x}h_1)$, respectively. The energy of incident electron is (**a**) $E = 56$ meV; (**b**) $E = 94$ meV; and (**c**) $E = 110$ meV. The other parameters are the same as those in Figure 2.

For the $V_3 < E < V_1$ case, we have $s_1 = s_4 = -1$, $s_2 = s_3 = 1$. Both Klein tunneling and classical motion are present in the double-well potential. Here, we choose $E = 94$ meV, and there are four intersections, as shown in Figure 3b. The corresponding wave functions of the four guided modes are presented in Figure 5: (a) $k_{2x}h_1 = 2.8959$; (b) $k_{2x}h_1 = 5.7775$; (c) $k_{2x}h_1 = 8.6220$; and (d) $k_{2x}h_1 = 11.3661$. The figure shows different mode structures and characteristics between Ψ_A and $-i\Psi_B$. For Ψ_A, the double-well potential can support the fundamental mode, first-order mode, second-order mode, and third-order mode. However, there is no fundamental mode in the waveguide for $-i\Psi_B$ (see Figure 5a, a small peak appears in wave functions for $-i\Psi_B$ on the left interface of the waveguide), and it only supports first-order mode, second-order mode, third-order mode, and fourth-order mode. This finding suggests that the electrons and holes have different behaviors under the same conditions, which is similar to the mixing case of classical motion and Klein tunneling that appeared in an asymmetric quantum well graphene waveguide [9,10].

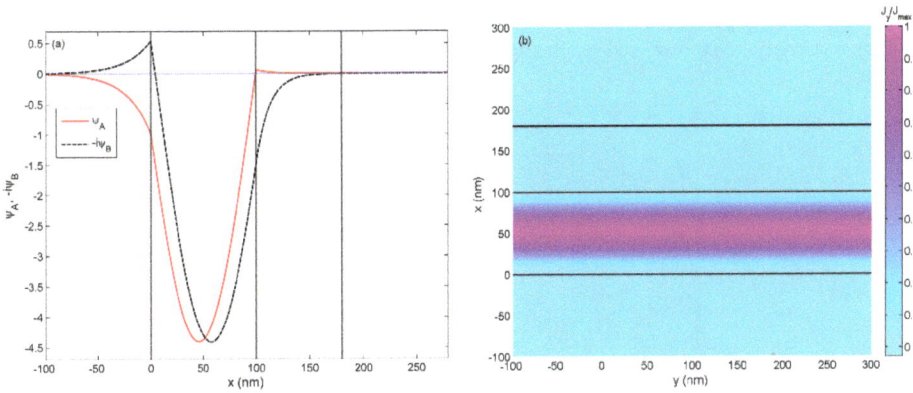

Figure 4. (**a**) the wave function of guided modes as a function of distance corresponding to the intersection ($k_{2x}h_1 = 2.9257, \theta = 69.8694°$) in Figure 3a. The solid curve and the dashed lines corresponds to Ψ_A and $-i\Psi_B$, respectively. The vertical lines represent the boundaries of the waveguide; and (**b**) the corresponding probability current density within the graphene waveguide for the guided mode. The solid black lines represent boundaries of the waveguide.

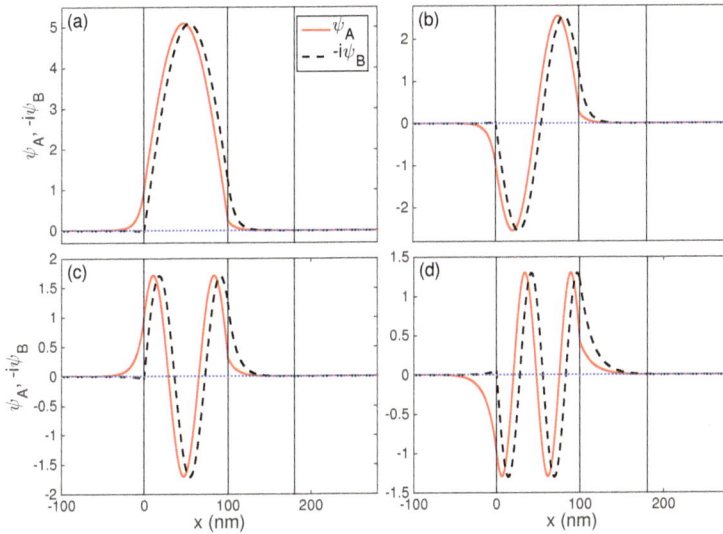

Figure 5. The wave function Ψ_A (solid) and $-i\Psi_B$ (dashed) of guided modes as a function of distance corresponding to the intersections in Figure 3b: (**a**) $k_{2x}h_1 = 2.8959$, $\theta = 78.2908°$; (**b**) $k_{2x}h_1 = 5.7775$, $\theta = 66.1160°$; (**c**) $k_{2x}h_1 = 8.6220$, $\theta = 52.8267°$; and (**d**) $k_{2x}h_1 = 11.3661$, $\theta = 37.1996°$.

For large electron energy $E = 110$ meV in the $E > V_1$ case, we have $s_1 = s_2 = s_3 = s_4 = 1$. The dependencies of $\tan(k_{2x}h_1)$ and $F(k_{2x}h_1)$ on $k_{2x}h_1$ are shown in Figure 3c. In this case, it is clearly seen from Figure 6 that the wave functions Ψ_A and $-i\Psi_B$ of guided modes for the five intersection points: (a) $k_{2x}h_1 = 2.9025$; (b) $k_{2x}h_1 = 5.7934$; (c) $k_{2x}h_1 = 8.6556$; (d) $k_{2x}h_1 = 11.4531$; and (e) $k_{2x}h_1 = 14.0545$ have very similar characteristics. The waveguide can support the fundamental mode, first-order mode, second-order mode, third-order mode, and fourth-order mode for the electrons and the holes. It must be pointed out that the guided modes in region 2 have similar characteristics with those in asymmetric quantum well graphene waveguide [9,10].

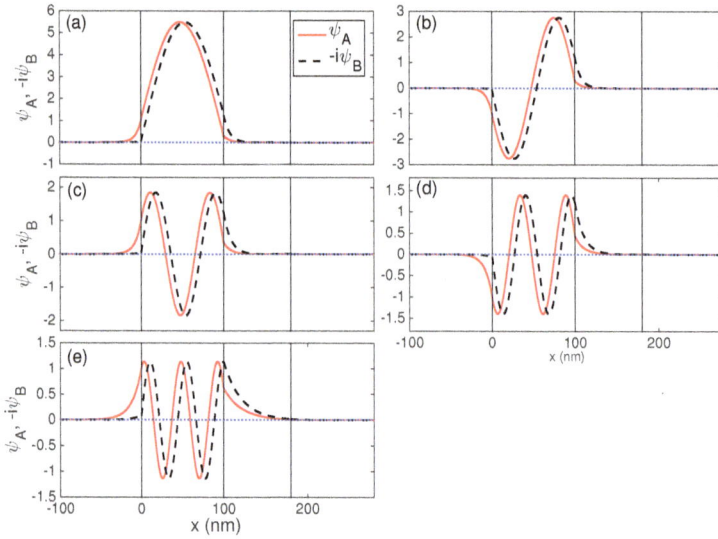

Figure 6. The wave function Ψ_A (solid) and $-i\Psi_B$ (dashed) of guided modes as a function of distance corresponding to the intersections in Figure 3c: (a) $k_{2x}h_1 = 2.9025$, $\theta = 79.9900°$; (b) $k_{2x}h_1 = 5.7934$, $\theta = 69.6994°$; (c) $k_{2x}h_1 = 8.6556$, $\theta = 58.7782°$; (d) $k_{2x}h_1 = 11.4531$, $\theta = 46.6948°$; and (e) $k_{2x}h_1 = 14.0545$, $\theta = 32.6828°$.

For the guided modes in the region of $0 < x < h_1 + h_2$, $k_{2x} = \sqrt{k_2^2 - k_y^2}$, $k_{3x} = \sqrt{k_3^2 - k_y^2}$ and $\alpha_1 = \alpha_4$ should be real. The transverse wavenumber k_{2x} in region 2 must fulfills the following condition:

$$\sqrt{k_2^2 - k_3^2} < k_{2x} < \sqrt{k_2^2 - k_1^2}. \tag{15}$$

For the incident electrons with low energy $E = 56$ meV, the double-well potential cannot support a guided mode. Thus, only a dispersion equation at higher energy $E = 94$ meV and $E = 110$ meV are plotted in Figure 7a,b, respectively. Each of them has three intersection points. For $E = 94$ meV (from Figure 7a), the intersections are (a) $k_{2x}h_1 = 13.4848$; (b) $k_{2x}h_1 = 14.0965$; and (c) $k_{2x}h_1 = 14.2552$, which are plotted in Figure 8. For $E = 110$ meV (from Figure 7b), we have (a) $k_{2x}h_1 = 15.2567$; (b) $k_{2x}h_1 = 16.0201$; and (c) $k_{2x}h_1 = 16.6082$, which are plotted in Figure 9.

From the figure, we see that the electrons and holes have similar mode structure and motion characteristics. For the $E = 94$ meV case, the double-well potential is able to support a higher order mode, such as the sixth-order mode (Figure 8b,c). This is known as mode double-degeneracy, which is similar to the oscillating guided modes in a symmetric five-layer left-handed waveguide [46]. For $E = 110$ meV, we have fifth-order mode, sixth-order mode, and seventh-order mode, as shown in Figure 9. For completeness, the probability current density of some specific guided modes is plotted in Figure 10, which shows good confinement.

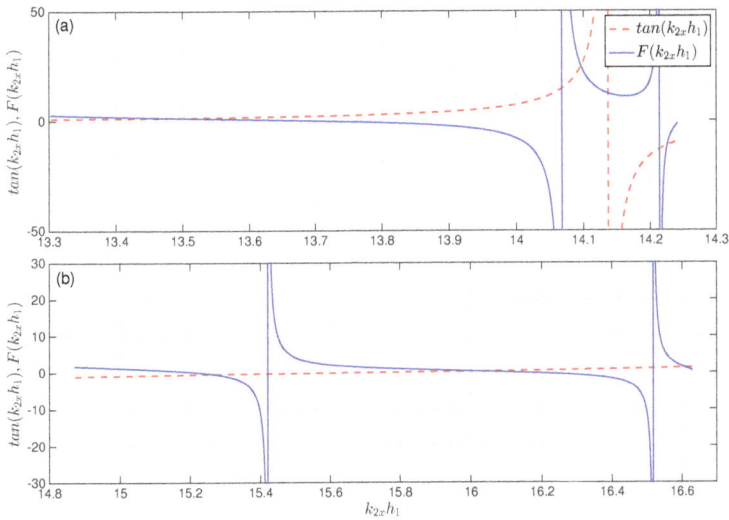

Figure 7. Graphical determination of $k_{2x}h_1$ for guided modes in $(0, h_1 + h_2)$. The **red** and **blue** curves correspond to $tan(k_{2x}h_1)$ and $F(k_{2x}h_1)$, respectively. The energy of incident electron is (**a**) $E = 94$ meV and (**b**) $E = 110$ meV. The other parameters are the same as those in Figure 2.

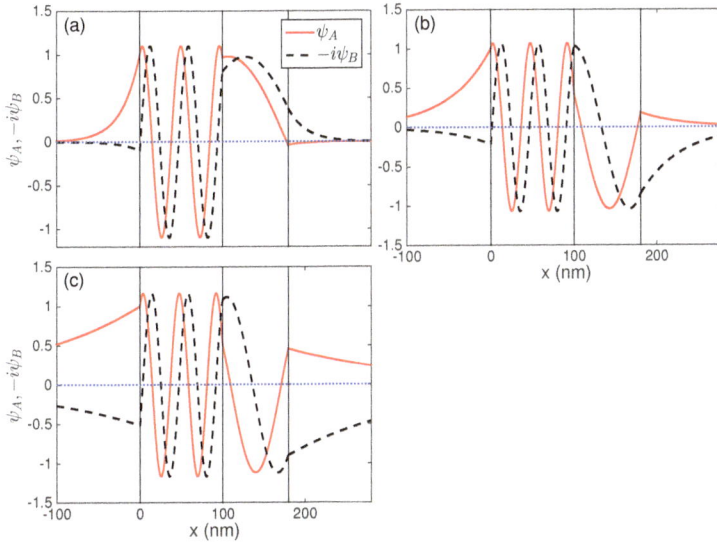

Figure 8. The wave function Ψ_A (solid) and $-i\Psi_B$ (dashed) of guided modes as a function of distance corresponding to the intersections in Figure 7a: (**a**) $k_{2x}h_1 = 13.4848$, $\theta = 19.0890°$; (**b**) $k_{2x}h_1 = 14.0965$, $\theta = 8.9296°$; and (**c**) $k_{2x}h_1 = 14.2552$, $\theta = 4.5134°$.

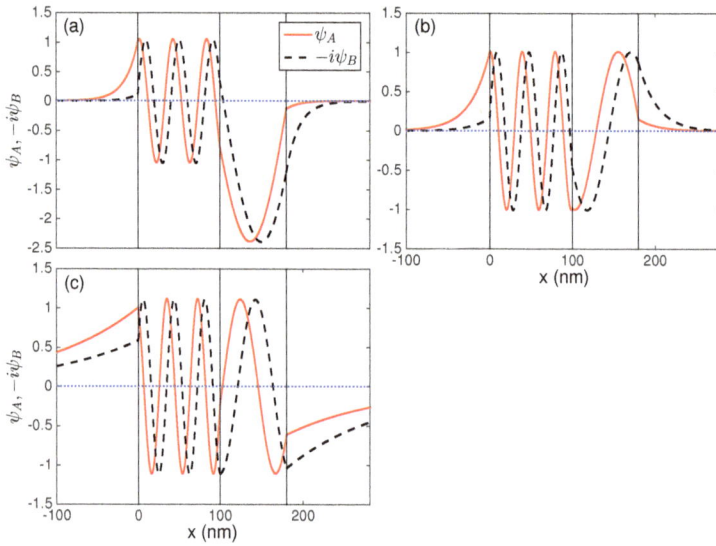

Figure 9. The wave function Ψ_A (solid) and $-i\Psi_B$ (dashed) of guided modes as a function of distance corresponding to the intersections in Figure 7b: (**a**) $k_{2x}h_1 = 15.2675$, $\theta = 23.8914°$; (**b**) $k_{2x}h_1 = 16.0201$, $\theta = 16.3885°$; and (**c**) $k_{2x}h_1 = 16.6082$, $\theta = 5.9544°$.

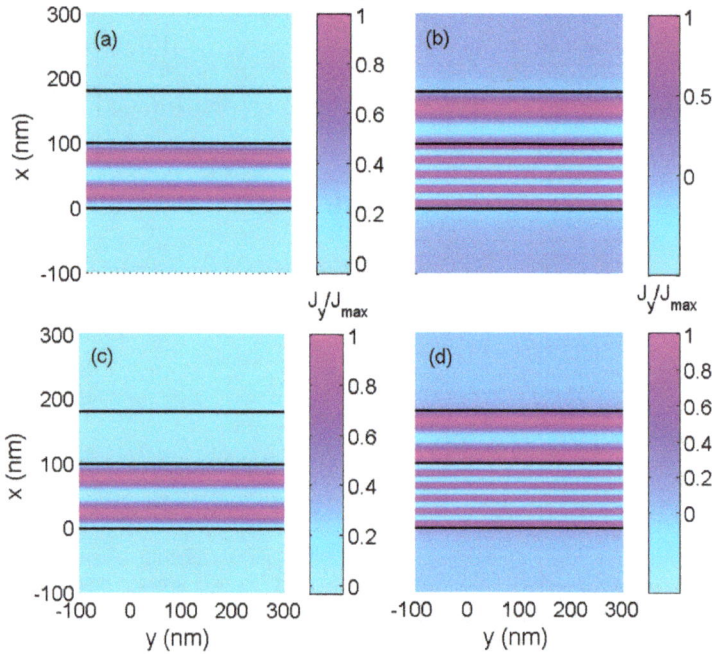

Figure 10. Probability current density within the graphene waveguide for the guided modes: (**a**) $E = 94$ meV, $k_{2x}h_1 = 5.7775$; (**b**) $E = 94$ meV, $k_{2x}h_1 = 14.0965$; (**c**) $E = 110$ meV, $k_{2x}h_1 = 5.7934$; and (**d**) $E = 110$ meV, $k_{2x}h_1 = 16.0201$.

In general, for the electrons in this graphene waveguide with $E = 94$ meV, the double-well potential supports the fundamental mode, first-order mode, second-order mode, third-order mode, fifth-order mode and sixth-order mode for Ψ_A, while it supports the first-order mode, second-order mode, third-order mode, fourth-order mode, fifth-order mode and sixth-order mode for $-i\Psi_B$. The fourth-order mode is absent for Ψ_A, while the fundamental mode is absent for $-i\Psi_B$. Mode double-degeneracy appears for the sixth-order mode for both Ψ_A and $-i\Psi_B$. The order of the guided modes is dependent on the incident energy and incidence angle for a given quantum well electron waveguide. The reason for the absence of some guided modes is that the incident energy is not sufficiently large with respect to the critical angle for the certain guided mode [8]. The absence of guided modes is similar to the situations in negative-refractive-index waveguides [46,47]. For the $E = 110$ meV case, the guided modes for Ψ_A and $-i\Psi_B$ have similar mode structure. The waveguide can support the fundamental mode, first-order mode, second-order mode, and up to the seventh-order mode for the electrons and the holes. There is no mode absent in this condition.

For many applications, it is desired to have mode tuning either by potential, the incident energy, or the well width. In Figure 11, the solutions to the guided modes at different values of well width $h_2 = 0, 40, 100, 160$ nm are presented. For the guided modes in $0 < x < h_1$, the results (see Figure 11a) show that changing h_2 has no effect on the number of guided modes. However, for the guided modes in $0 < x < h_1 + h_2$, we have more guide modes at higher h_2, as shown in Figure 11b. For example, at $h_2 = 160$ nm, we have three guided modes in the region $0 < x < h_1 + h_2$, which is higher than two modes for $h_2 = 100$ nm, and only one mode for $h_2 = 40$ nm. Obviously, there are no intersections for guided modes in the region $0 < x < h_1 + h_2$ with $h_2 = 0$ nm. In comparison with the quantum well graphene waveguide, the double-well graphene waveguide can support some higher order guided modes in a wider range.

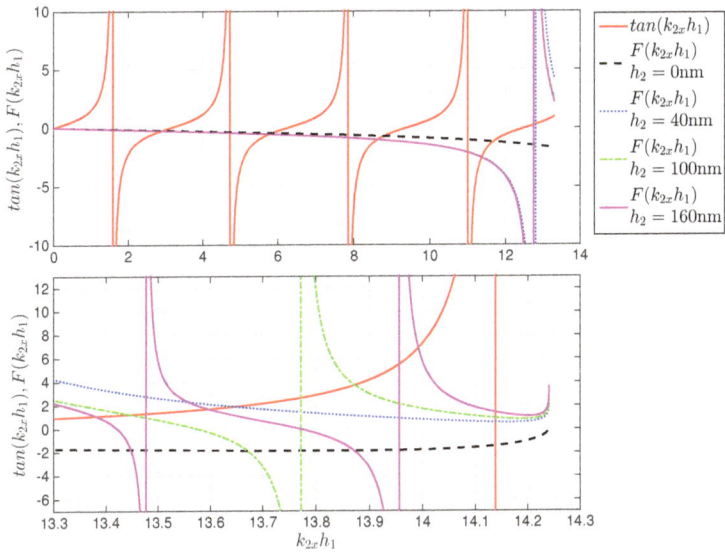

Figure 11. Graphical determination of $\tan(k_{2x}h_1)$ and $F(k_{2x}h_1)$ for guided modes in (**a**) $(0, h_1)$ and (**b**) $(0, h_1 + h_2)$ with different values of h_2. Here, the incident energy of electrons is fixed at $E = 94$ meV. The other parameters are the same as those in Figure 2.

4. Conclusions

We have determined the guided modes in a double-well asymmetry potential with a transfer matrix method over a wide range of parameters. Two types of guide modes were found in the proposed structures, one of which was similar to the characteristics with those in the graphene waveguide with asymmetric quantum well structure. It was found that the proposed graphene based electron waveguides can support some higher order nodes (up to the 7th order). Mode double-degeneracy appeared under some conditions. For some given incident energy of electrons, some modes are absent. Furthermore, the tuning of the number of guided modes by the well width was also studied. These novel properties of the guided modes in the double-well potential may provide potential applications for various graphene-based electronic devices. Electron waveguides are examples of various graphene-based electronic devices; however, the transfer matrix method may not be suitable for simulating other graphene-based electronic devices, such as graphene-based field effect transistors. Hence, we will seek other methods to solve the Dirac equation in graphene in order to simulate graphene-based electronic devices in our future research work.

Acknowledgments: This work is supported by the Singapore Ministry of Education T2 Grant No. T2MOE1401 and the U.S. Air Force Office of Scientific Research (AFOSR) through the Asian Office of Aerospace Research and Development (AOARD) under Grant No. FA2386-14-1-4020.

Author Contributions: Yi Xu and L.K. Ang proposed the idea and presided over the study. Yi Xu and L.K. Ang conceived and calculated the setup. Yi Xu and L.K. Ang wrote the paper. All authors read and approved the final manuscript.

Conflicts of Interest: The authors declare no conflict of interest.

References

1. Novoselov, K.S.; Geim, A.K.; Morozov, S.V.; Jiang, D.; Zhang, Y.; Dubonos, S.V.; Grigorieva, I.V.; Firsov, A.A. Electric field effect in atomically thin carbon films. *Science* **2004**, *306*, 666–669.
2. Beenakker, C.; Sepkhanov, R.; Akhmerov, A.; Tworzydło, J. Quantum Goos-Hänchen effect in graphene. *Phys. Rev. Lett.* **2009**, *102*, 146804.
3. Sharma, M.; Ghosh, S. Electron transport and Goos–Hänchen shift in graphene with electric and magnetic barriers: Optical analogy and band structure. *J. Phys. Condens. Matter* **2011**, *23*, 055501.
4. Cheianov, V.V.; Fal'ko, V.; Altshuler, B. The focusing of electron flow and a Veselago lens in graphene pn junctions. *Science* **2007**, *315*, 1252–1255.
5. Park, C.H.; Son, Y.W.; Yang, L.; Cohen, M.L.; Louie, S.G. Electron beam supercollimation in graphene superlattices. *Nano Lett.* **2008**, *8*, 2920–2924.
6. Asmar, M.M.; Ulloa, S.E. Rashba spin-orbit interaction and birefringent electron optics in graphene. *Phys. Rev. B* **2013**, *87*, 075420.
7. Ghosh, S.; Sharma, M. Electron optics with magnetic vector potential barriers in graphene. *J. Phys. Condens. Matter* **2009**, *21*, 292204.
8. Zhang, F.M.; He, Y.; Chen, X. Guided modes in graphene waveguides. *Appl. Phys. Lett.* **2009**, *94*, 212105.
9. He, Y.; Xu, Y.; Yang, Y.F.; Huang, W.D. Guided modes in asymmetric graphene waveguides. *Appl. Phys. A* **2014**, *115*, 895–902.
10. Ping, P.; Peng, Z.; Liu, J.K.; Cao, Z.Z.; Li, G.Q. Oscillating guided modes in graphene-based asymmetric waveguides. *Commun. Theor. Phys.* **2012**, *58*, 765.
11. Xu, Y.; Ang, L.K. Guided modes in a triple-well graphene waveguide: Analogy of five-layer optical waveguide. *J. Opt.* **2015**, *17*, 035005.
12. Williams, J.; Low, T.; Lundstrom, M.; Marcus, C. Gate-controlled guiding of electrons in graphene. *Nat. Nanotechnol.* **2011**, *6*, 222–225.
13. Myoung, N.; Ihm, G.; Lee, S. Magnetically induced waveguide in graphene. *Phys. Rev. B* **2011**, *83*, 113407.
14. Huang, W.D.; He, Y.; Yang, Y.F.; Li, C.F. Graphene waveguide induced by gradually varied magnetic fields. *J. Appl. Phys.* **2012**, *111*, 053712.
15. Hartmann, R.R.; Robinson, N.; Portnoi, M. Smooth electron waveguides in graphene. *Phys. Rev. B* **2010**, *81*, 245431.

16. Wu, Z.H.; Zhai, F.; Peeters, F.; Xu, H.; Chang, K. Valley-dependent Brewster angles and Goos-Hänchen effect in strained graphene. *Phys. Rev. Lett.* **2011**, *106*, 176802.
17. Villegas, C.E.; Tavares, M.R.; Hai, G.Q.; Peeters, F. Sorting the modes contributing to guidance in strain-induced graphene waveguides. *New J. Phys.* **2013**, *15*, 023015.
18. Pereira, V.M.; Neto, A.C. Strain engineering of graphene's electronic structure. *Phys. Rev. Lett.* **2009**, *103*, 046801.
19. Wu, Z.H. Electronic fiber in graphene. *Appl. Phys. Lett.* **2011**, *98*, 082117.
20. Pereira, J.M., Jr.; Mlinar, V.; Peeters, F.; Vasilopoulos, P. Confined states and direction-dependent transmission in graphene quantum wells. *Phys. Rev. B* **2006**, *74*, 045424.
21. Tudorovskiy, T.Y.; Chaplik, A. Spatially inhomogeneous states of charge carriers in graphene. *JETP Lett.* **2007**, *84*, 619–623.
22. De Martino, A.; Dell'Anna, L.; Egger, R. Magnetic confinement of massless Dirac fermions in graphene. *Phys. Rev. Lett.* **2007**, *98*, 066802.
23. Low, T.; Guinea, F. Strain-induced pseudomagnetic field for novel graphene electronics. *Nano Lett.* **2010**, *10*, 3551–3554.
24. Elton, D.M.; Levitin, M.; Polterovich, I. *Eigenvalues of a One-Dimensional Dirac Operator Pencil*; Annales Henri Poincaré; Springer: Berlin, Germany, 2014; Volume 15, pp. 2321–2377.
25. Hartmann, R.R.; Portnoi, M. Quasi-exact solution to the Dirac equation for the hyperbolic-secant potential. *Phys. Rev. A* **2014**, *89*, 012101.
26. Stone, D.; Downing, C.; Portnoi, M. Searching for confined modes in graphene channels: The variable phase method. *Phys. Rev. B* **2012**, *86*, 075464.
27. Yuan, J.H.; Cheng, Z.; Zeng, Q.J.; Zhang, J.P.; Zhang, J.J. Velocity-controlled guiding of electron in graphene: Analogy of optical waveguides. *J. Appl. Phys.* **2011**, *110*, 103706.
28. Li, H.D.; Wang, L.; Lan, Z.H.; Zheng, Y.S. Generalized transfer matrix theory of electronic transport through a graphene waveguide. *Phys. Rev. B* **2009**, *79*, 155429.
29. Pedersen, J.G.; Gunst, T.; Markussen, T.; Pedersen, T.G. Graphene antidot lattice waveguides. *Phys. Rev. B* **2012**, *86*, 245410.
30. Allen, M.T.; Shtanko, O.; Fulga, I.C.; Akhmerov, A.; Watanabe, K.; Taniguchi, T.; Jarillo-Herrero, P.; Levitov, L.S.; Yacoby, A. Spatially resolved edge currents and guided-wave electronic states in graphene. *Nat. Phys.* **2016**, *12*, 128–133.
31. Wang, L.G.; Zhu, S.Y. Electronic band gaps and transport properties in graphene superlattices with one-dimensional periodic potentials of square barriers. *Phys. Rev. B* **2010**, *81*, 205444.
32. Xu, Y.; He, Y.; Yang, Y.F. Transmission gaps in graphene superlattices with periodic potential patterns. *Phys. B Condens. Matter* **2015**, *457*, 188–193.
33. Xu, Y.; He, Y.; Yang, Y.F.; Zhang, H.F. Electronic band gaps and transport in Cantor graphene superlattices. *Superlatt. Microstruct.* **2015**, *80*, 63–71.
34. Hanson, G.W. Dyadic Green's functions and guided surface waves for a surface conductivity model of graphene. *J. Appl. Phys.* **2008**, *103*, 064302.
35. Tamagnone, M.; Gomez-Diaz, J.; Mosig, J.R.; Perruisseau-Carrier, J. Reconfigurable terahertz plasmonic antenna concept using a graphene stack. *Appl. Phys. Lett.* **2012**, *101*, 214102.
36. Bouzianas, G.; Kantartzis, N.; Tsiboukis, T. Subcell dispersive finite-difference time-domain schemes for infinite graphene-based structures. *IET Microw. Antennas Propag.* **2012**, *6*, 377–386.
37. Llatser, I.; Kremers, C.; Cabellos-Aparicio, A.; Jornet, J.M.; Alarcón, E.; Chigrin, D.N. Graphene-based nano-patch antenna for terahertz radiation. *Photonics Nanostruct. Fundam. Appl.* **2012**, *10*, 353–358.
38. Salonikios, V.; Amanatiadis, S.; Kantartzis, N.; Yioultsis, T. Modal analysis of graphene microtubes utilizing a two-dimensional vectorial finite element method. *Appl. Phys. A* **2016**, *122*, 1–7.
39. Nayyeri, V.; Soleimani, M.; Ramahi, O.M. Modeling graphene in the finite-difference time-domain method using a surface boundary condition. *IEEE Trans. Antennas Propag.* **2013**, *61*, 4176–4182.
40. Amanatiadis, S.A.; Kantartzis, N.V.; Tsiboukis, T.D. A loss-controllable absorbing boundary condition for surface plasmon polaritons propagating onto graphene. *IEEE Trans. Mag.* **2015**, *51*, 1–4.
41. Mock, A. Padé approximant spectral fit for FDTD simulation of graphene in the near infrared. *Opt. Mater. Express* **2012**, *2*, 771–781.

42. Huard, B.; Sulpizio, J.; Stander, N.; Todd, K.; Yang, B.; Goldhaber-Gordon, D. Transport measurements across a tunable potential barrier in graphene. *Phys. Rev. Lett.* **2007**, *98*, 236803.

43. Brey, L.; Fertig, H. Electronic states of graphene nanoribbons studied with the Dirac equation. *Phys. Rev. B* **2006**, *73*, 235411.

44. Mhamdi, A.; Salem, E.B.; Jaziri, S. Electronic reflection for a single-layer graphene quantum well. *Solid State Commun.* **2013**, *175*, 106–113.

45. Allain, P.E.; Fuchs, J. Klein tunneling in graphene: Optics with massless electrons. *Eur. Phys. J. B* **2011**, *83*, 301–317.

46. He, Y.; Zhang, J.; Li, C.F. Guided modes in a symmetric five-layer left-handed waveguide. *J. Opt. Soc. Am. B* **2008**, *25*, 2081–2091.

47. Shadrivov, I.V.; Sukhorukov, A.A.; Kivshar, Y.S. Guided modes in negative-refractive-index waveguides. *Phys. Rev. E* **2003**, *67*, 057602.

electronics

MDPI

Article

Modeling and Design of a New Flexible Graphene-on-Silicon Schottky Junction Solar Cell

Francesco Dell'Olio, Michele Palmitessa and Caterina Ciminelli *

Optoelectronics Laboratory, Politecnico di Bari, via E. Orabona 4, 70125 Bari, Italy;
francesco.dellolio@poliba.it (F.D.O.); mipalm@fastwebnet.it (M.P.)
* Correspondence: caterina.ciminelli@poliba.it; Tel.: +39-080-596-3404

Academic Editors: Yoke Khin Yap and Zhixian Zhou
Received: 5 September 2016; Accepted: 19 October 2016; Published: 26 October 2016

Abstract: A new graphene-based flexible solar cell with a power conversion efficiency >10% has been designed. The environmental stability and the low complexity of the fabrication process are the two main advantages of the proposed device with respect to other flexible solar cells. The designed solar cell is a graphene/silicon Schottky junction whose performance has been enhanced by a graphene oxide layer deposited on the graphene sheet. The effect of the graphene oxide is to dope the graphene and to act as anti-reflection coating. A silicon dioxide ultrathin layer interposed between the n-Si and the graphene increases the open-circuit voltage of the cell. The solar cell optimization has been achieved through a mathematical model, which has been validated by using experimental data reported in literature. The new flexible photovoltaic device can be integrated in a wide range of microsystems powered by solar energy.

Keywords: optoelectronic devices; solar cell; Schottky junction; flexible electronics; graphene

---

1. Introduction

Graphene, which is one of the most promising two-dimensional (2D) materials, is composed of carbon atoms arranged in a hexagonal lattice [1]. It can be synthetized in the form of ultrathin sheets consisting of one or a few atomic layers via several techniques, such as chemical vapor deposition, and it can be easily transferred on different substrates [2].

Graphene has several unique electronic, optoelectronic, mechanical, and thermal properties that make it very attractive in many scientific areas, including nanoelectronics, optoelectronics, and photonics [3].

The research interest on graphene optoelectronics and photonics [4] is quickly growing with the demonstration of high-performing devices such as modulators [5], photodetectors [6], saturable absorbers [7], absorbers in the terahertz regime [8,9], polarization controllers [10], delay lines [11,12], phase shifters [13], and solar cells [14]. In particular, the large mechanical flexibility of graphene, its high conductivity (10^6 S/cm), and transparency (97.7% for graphene monolayer in visible wavelengths) make the use of this material very attractive in the field of photovoltaics [15].

Several enabling operative functions have been envisaged for graphene sheets in photovoltaics technology. Graphene can serve as a transparent conductive electrode in organic or inorganic solar cells [16,17], as an intermediate layer in tandem solar cells [18], and as a barrier layer in perovskite solar cells [19]. Perovskite solar cells in which the electron collection layers are implemented through especially synthetized graphene/TiO_2 nanocomposites have been demonstrated [20], while graphene quantum dots have been used as active layer in a solar cell [21].

By transferring a graphene sheet on a semiconducting substrate of Si or GaAs, a Shottky junction acting as a solar cell under illumination can be manufactured via simple fabrication processes.

The first graphene/n-Si Shottky junction solar cell exhibited a quite low efficiency (~1.5%) [22]. By properly optimizing the cell configuration, the efficiency has been improved up to 15.6% [23]. A more expensive graphene/GaAs Shottky junction solar cell with an efficiency of >18% has been reported [24]. These values of efficiency are the best ones for graphene-based solar cells.

One of the key advantages of the graphene/n-Si Shottky junction solar cells is that they can be made flexible by thinning the substrate [25,26]. In this way, flexible solar cells with an efficiency potentially higher than 10% and an excellent environmental stability can be fabricated via a simple and low cost technological process. This feature of the graphene/n-Si Shottky junction solar cells is extremely attractive because flexible solar cells [27] are in demand for a wide range of applications, such as wearable and implantable microsystems and wireless sensor networks for Internet of Things (IoT).

High-efficiency flexible solar cells with good environmental stability, low complexity, and a low manufacturing have not yet been proposed. The state of the art of the efficiency for flexible solar cells is around 30% for GaAs cells manufactured via epitaxial lift-off [28]. However, the only solar cells with an efficiency >10% that can be easily fabricated via a low cost process are the perovskite-based cells [29] (record efficiency = 16.47%), even if at the expense of poor environmental stability [30].

In this paper, we report on the design of a new flexible solar cell based on a Schottky junction consisting of an ultrathin layer of graphene oxide (GO) deposited on a few graphene atomic layers, which are transferred on an n-doped silicon layer (see Figure 1).

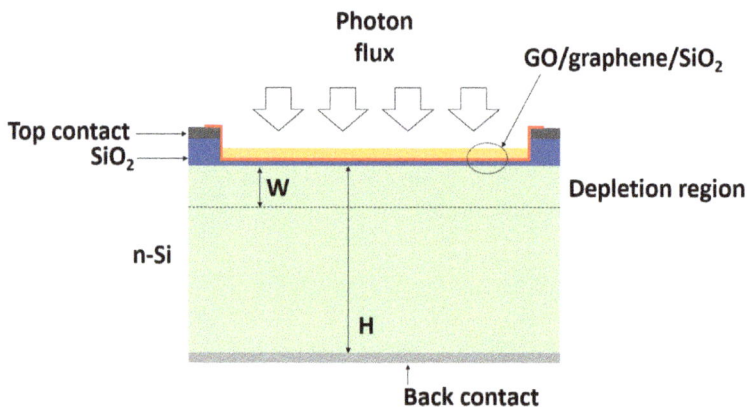

Figure 1. Cross section of the designed Schottky junction solar cell. GO: graphene oxide.

Aiming at maximizing the cell performance, the presence of a silicon oxide layer (thickness <2 nm) between the silicon layer and the graphene sheet has been considered, and the influence of a graphene oxide film (thickness <100 nm) on the top of the graphene sheet has been studied.

The silicon dioxide, with an optimized thickness, increases the open-circuit voltage and consequently the efficiency of the solar cell. The carrier transport across the silicon dioxide ultrathin layer is enabled by the tunnel effect.

The graphene oxide film induces the p-doping of the graphene [31] with a consequent increase of both the work function and the conductivity of the graphene sheet. In addition, the GO film acts as an anti-reflection coating layer, as explained in Section 3.

We assume that the graphene sheet, synthetized by chemical vapor deposition, is transferred on the top of a square Si window patterned in a Si/SiO$_2$ wafer that is properly thinned to guarantee the desired flexibility. In addition, we assume that, before the graphene transfer, a metal contact (top contact) with a square shape is deposited on the SiO$_2$ layer around the Si window and another metal contact (back contact) is deposited on the back side of the n-Si. The ultrathin SiO$_2$ layer can be grown on the n-Si immediately before the graphene transfer by exposing the wafer to clean room air at room temperature and controlled humidity (average humidity = 42%). In this way, the thickness of the SiO$_2$ layer slowly increases as the exposure time increases [32]. Alternatively, the ultrathin SiO$_2$ layer can be grown by immersing the sample in ultrapure water with dissolved oxygen concentration of 9 ppm. For example, by using this technique, a SiO$_2$ layer with a thickness of 1.5 nm can be grown of an n-doped silicon wafer (donor dopant concentration = 10^{15} cm^{-3}) after an exposure time of 2×10^4 min [32]. After the graphene transfer, a GO solution can be spin-coated on the cell forming a uniform GO layer.

The solar cell design is based on a mathematical model taking into account all physical effects occurring within the Schottky junction when it is illuminated via solar radiation. The model is validated using data from the literature. The device performance has been calculated assuming AM (air mass) 1.5 illumination (power density = 100 mW/cm^2), which is the standard terrestrial spectrum of solar radiation.

2. Solar Cell Model

The complete mathematical model was developed for the design of the solar cell shown in Figure 1. Simpler structures without the GO and the SiO$_2$ layers can be also studied with the same model.

As is well known, a depletion layer is created in the n-Si in close proximity to the junction. Photons arriving within the silicon layer generate electron-hole pairs both in the depletion layer and outside of it. All of these pairs can contribute to the photo-generated current density J_{ph}.

J_{ph} originates from two contributions—the drift current density J_{dr} and the hole current density J_p. J_{dr} is due to electrons and holes that are generated in the depletion layer by the shorter wavelength light and then accelerated by the built-in electric field towards the metal contacts before they recombine. J_p is due to the holes generated outside the depletion layer by the longer wavelength light. The holes diffuse in the n-Si towards the top contact.

According to [33], the wavelength dependent expressions of J_{dr} and J_p are given by

$$J_{dr}(\lambda) = qF(\lambda) T_R(\lambda) \left[1 - e^{-\alpha(\lambda)W}\right], \text{ and} \tag{1}$$

$$J_p(\lambda) = \frac{qF(\lambda)T_R(\lambda)\alpha(\lambda)L_p}{\alpha^2(\lambda)L_p^2 - 1} e^{-\alpha(\lambda)W} \left[\alpha(\lambda)L_p - \frac{\frac{SL_p}{D_p}[\cosh\frac{H'}{L_p} - e^{-\alpha(\lambda)H'}] + \sinh\frac{H'}{L_p} + \alpha(\lambda)L_p e^{-\alpha(\lambda)H'}}{\frac{SL_p}{D_p}\sinh\frac{H'}{L_p} + \cosh\frac{H'}{L_p}}\right], \tag{2}$$

where λ is the wavelength, q is the electron charge, F is the photon flux, α is the silicon absorption coefficient, W is the width of the depletion layer, S is the recombination velocity at the back contact, and H' is equal to $H-W$ where H is the thickness of the n-Si layer. L_p is the hole diffusion length given by $(D_p \tau_p)^{\frac{1}{2}}$ where D_p is the hole diffusion coefficient and τ_p is the hole lifetime.

$T_R(\lambda)$ is given by $T_{ML}(\lambda)[1-R(\lambda)]$; $T_{ML}(\lambda)$ is the transmittance of the GO/graphene/SiO$_2$ multi-layer, and R is the reflectance of the GO-coated graphene/silicon junction. When the cell structure does not include the GO and SiO$_2$ layers, $T_{ML}(\lambda)$ denotes the transmittance of the graphene sheet, and R the silicon reflectance.

The width W of the depletion layer can be written as [33]

$$W = \left\{\frac{2\varepsilon_s}{qN_D}\left[(\Phi_G - \chi) - \frac{kT \ln(N_C/N_D)}{q} - \frac{kT}{q}\right]\right\}^{1/2}, \tag{3}$$

where ε_s is the silicon dielectric constant, N_D is the donor impurity concentration, Φ_G is the graphene work function, χ is the silicon electron affinity, k is the Boltzmann constant, T is the absolute temperature, and N_C is the effective density of states in the conduction band.

For each value of λ, $J_{ph}(\lambda) = J_{dr}(\lambda) + J_p(\lambda)$. Since silicon absorbs light in the wavelength range from $\lambda_1 = 0.28$ μm to $\lambda_2 = 1.2$ μm, the total photo-generated current is given by the following integral:

$$J_{ph} = \int_{\lambda_1}^{\lambda_2} \left[J_{dr}(\lambda) + J_p(\lambda) \right]. \tag{4}$$

The external quantum efficiency of the cell (EQE), i.e., the number of electron-hole pairs generated for each absorbed photon, is given by

$$EQE(\lambda) = \frac{J_{ph}(\lambda)}{qF(\lambda)\,T_R(\lambda)}. \tag{5}$$

Assuming that the shunt resistance is infinite (see, for example, the experimental data in [22]) and, consequently, its effect on the performance of the cell is negligible, the equivalent electric circuit in Figure 2a models the solar cell.

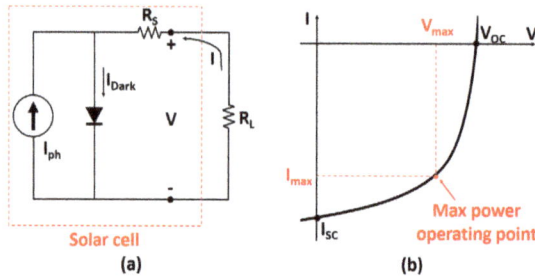

Figure 2. (a) Equivalent circuit of the solar cell. (b) Qualitative plot of the I–V characteristics of the solar cell.

The current generator represents the photocurrent generation and $I_{ph} = J_{ph}\,A$, where A is the area of the solar cell. The resistance R_S is the series resistance of the solar cell. The diode corresponds to the cell behavior when it is not illuminated by the solar radiation (under this condition $I_{ph} = 0$). The current flowing in the diode is the dark current I_{Dark} equal to

$$I_{Dark} = I_S \left(e^{\frac{q(V - IR_S)}{nkT}} - 1 \right), \tag{6}$$

where n is the ideality factor, and I_S is the saturation current. I_S can be written as [33,34]

$$I_S = AA^* T^2 e^{\frac{-q(\Phi_G - \chi)}{kT}} e^{-\delta\sqrt{q\Phi_T}}, \tag{7}$$

where A^* is the effective Richardson constant, ϕ_T (in eV) is the average height of the energy barrier due to the SiO$_2$ layer, and δ is the thickness of the SiO$_2$ layer ($\delta = 0$ when there is not a SiO$_2$ layer interposed between the graphene sheet and the n-Si).

The I–V characteristics (see Figure 2b) of the solar cell is

$$I = I_S \left(e^{\frac{q(V - IR_S)}{nkT}} - 1 \right) - I_{ph}. \tag{8}$$

V_{OC} denotes the open-circuit voltage, which is the voltage across the cell terminals when the load resistance R_L is an open circuit and, consequently, $I = 0$, and I_{SC} denotes the short-circuit current, i.e., the value of I when $R_L = 0$ and, consequently, $V = 0$.

The fill factor FF of the solar cell, which is a measure of the "squareness" of the I–V characteristics, is given by:

$$FF = \frac{V_{max}\,J_{max}}{V_{OC}\,J_{SC}}, \tag{9}$$

where V_{max} and J_{max} are the voltage and the current density at the maximum-power operating point, respectively, i.e., the point of the I–V characteristics of the solar cell where the generated electric power $P = V I$ is maximum.

The power conversion efficiency (PCE; sometimes called simply efficiency) is the main performance parameter for any solar cell:

$$PCE = \frac{V_{max}\,J_{max}}{P_{in}}, \tag{10}$$

where P_{in} is the incident optical power density.

The model physical parameters used in the simulations are summarized in Table 1. The wavelength-dependent silicon absorption coefficient was derived from experimental data [35] and is shown in Figure 3. The empirical relation reported in [36] was used to take into account the τ_p dependence on N_D. D_p was expressed as $kT\,(\mu_p/q)$, and the analytical expression in [37] was used for the hole mobility μ_p. Data in [31] were used to estimate R_S, the reflectivity $R(\lambda)$, and the wavelength-dependent transmittance of graphene, GO, and SiO_2.

Table 1. Model parameters.

Parameter	Symbol	Value	Reference
Temperature	T	300 K	[34]
Electron affinity of silicon	χ	4.05 eV	[34]
Electron charge	q	1.6×10^{-19} C	[34]
Effective density of states in conduction band	N_C	2.8×10^{19} cm^{-3}	[34]
Ideality factor	n	1.42 (with GO layer)/1.65 (without GO layer)	[31]
Graphene work function	Φ_G	4.92 eV (with GO layer)/4.84 eV (without GO layer)	[31]
Effective Richardson constant	A^*	112 A·cm^{-2}·K^{-2}	[31]
Silicon dielectric constant	ε_s	11.7	[34]
Recombination velocity at the back contact	S	10^{15} cm/s	[38]
Boltzmann constant	k	8.62 eV/K	[34]

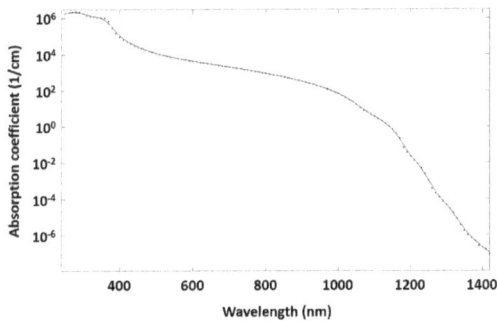

Figure 3. Wavelength-dependent silicon absorption coefficient.

The model was used to calculate the main performance parameters of two graphene-based Schottky junction solar cells reported in the literature. The calculated performance parameters were then compared to the experimental values in the literature, showing a very good model/experiment agreement.

The first device considered for the validation of our model is a solar cell made by a monolayer graphene/ultrathin silicon Schottky junction [25]. The silicon layer is n-doped with $N_D = 2 \times 10^{16}$ cm^{-3} and its thickness is only 10.6 μm. We calculated the J–V curve and the EQE dependence on the wavelength. By comparing the plot obtained by the model with the experimental curves, we noticed a very good agreement, as described below.

Table 2 compares the calculated and measured values of the four key performance parameters, V_{OS}, J_{SC}, FF, and PCE. For the PCE, the difference between calculated and measured values is about 0.3%. For V_{OS}, J_{SC}, and FF the differences are 0.03 V, 1.11 mA/cm^2, and 3.3%, respectively.

Table 2. Calculated vs. measured performance parameters of the solar cell reported in [25].

Measured/Calculated	V_{OC} (V)	J_{SC} (mA/cm^2)	FF (%)	PCE (%)
Measured values [25]	0.416	12.40	25.2	1.30
Calculated values	0.419	13.51	28.5	1.61

The second solar cell simulated by our model is that reported in [31]. It is a cell based on a monolayer graphene/n-Si Schottky junction that has been experimentally studied with and without a GO layer with a thickness of 100 nm on the graphene sheet. The thickness of the silicon layer is 300 μm, and the dopant atoms concentration in this layer is $N_D = 5 \times 10^{15}$ cm^{-3}. Using our model, we simulated the two cells with and without the GO layer. For example, Figure 4a shows the calculated J–V curve for the two cells. Again, the agreement between the calculated (Figure 4a) and the measured curves (Figure 4b) is very good. The same agreement can be observed for the performance parameters (see Table 3) with a difference between calculated and measured values of the PCE less than 0.3%. For V_{OS}, J_{SC}, and FF the differences are <0.015 V, <5 mA/cm^2, and ≤5%, respectively.

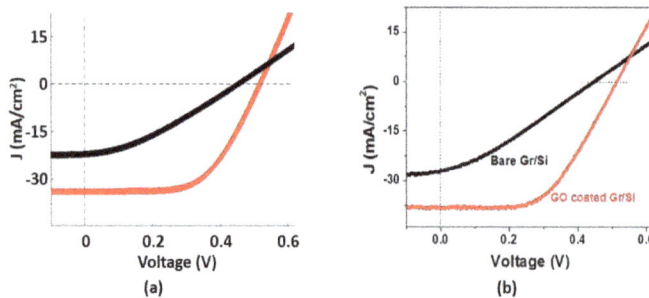

Figure 4. J–V curve for the solar cell in [31]. Black curves refer to the cell w/o graphene oxide (GO). Red curves refer to the cell with GO. (a) Calculated data. (b) Measured data that are adapted from [31], with the permission of The Royal Society of Chemistry.

Table 3. Calculated vs. measured performance parameters of the solar cell reported in [31].

Measured/Calculated	V_{OC} (V)	J_{SC} (mA/cm^2)	FF (%)	PCE (%)
Measured values (w/o GO) [31]	0.440	27.20	29	3.6
Calculated values (w/o GO)	0.453	22.49	33.14	3.38
Measured values (with GO) [31]	0.512	38.40	53	10.6
Calculated values (with GO)	0.518	34.10	58	10.33

3. Solar Cell Design

Aiming to investigate the influence of silicon doping on solar cell performance, we considered a simple monolayer graphene/n-Si junction with a thickness $H = 10$ μm. This H value assures a very good flexibility of the cell. We varied N_D from 10^{15} to 10^{19} cm^{-3}, and we calculated the relevant values of the PCE (see Figure 5a). As expected, the PCE increases as N_D decreases because both the photo-generated current density J_{ph} and, consequently, the EQE increases when N_D decreases. Due to this PCE dependence on N_D, we chose a light doping of the n-Si layer with $N_D = 10^{15}$ cm^{-3}.

The number of atomic layers forming the graphene sheet has to be optimized because the graphene sheet resistance and, consequently, R_S decrease as the number of atomic layers increases. When R_S decrease, an improvement of the efficiency results, but the graphene optical transmittance and thus J_{ph} are reduced when the number of atomic layers is too high. Considering a graphene/n-Si solar cell with silicon doping $N_D = 10^{15}$ cm^{-3} and $H = 10$ μm, we evaluated the PCE dependence on the number of atomic layers when this parameter varies from 1 to 5 (see Figure 5b). The best PCE value (=3.02%) was obtained when the number of atomic layers was 3. This result is consistent with experimental results reported in [26].

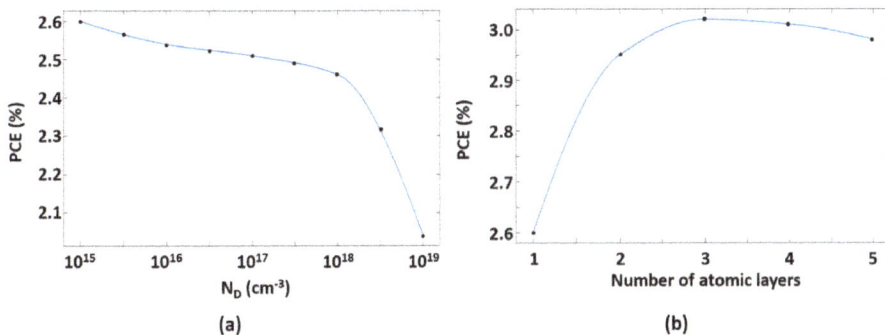

Figure 5. (**a**) Power conversion efficiency (PCE) dependence on N_D for the monolayer graphene/n-Si junction with $H = 10$ μm. (**b**) PCE dependence on the number of graphene atomic layers forming the graphene sheet, $H = 10$ μm and $N_D = 10^{15}$ cm^{-3}.

To further increase the PCE of the cell, a quarter-wave anti-reflection coating (ARC) layer can be deposited on the trilayer-graphene. The optimum refractive index of the ARC layer n_{ARC} is the geometric mean of the refractive index of silicon and air [39]. Since the solar energy intensity is maximum at about 0.55 μm [40] and the silicon refractive index at that wavelength is 4.1 [35], the optimum value of n_{ARC} is 2.0, which is very close to the refractive index of a GO layer with a thickness of <100 nm [31]. Thus, a GO layer can be used as ARC. As already mentioned, that layer dopes the graphene sheet, with a further beneficial effect on the cell performance. The optimum thickness of the ARC layer is $\lambda/4n_{ARC} = 69$ nm at $\lambda = 0.55$ μm. Table 4 summarizes the solar cell performance with and without the GO layer and shows that the PCE improves up to 5.31% when the GO is deposited. V_{OS}, J_{SC}, and FF increase due to the deposition of the GO layer of 0.071 V, 5.35 mA/cm^2, and 3.86%, respectively.

Table 4. Performance of the solar cell with and without the GO quarter-wave anti-reflection coating (ARC).

Solar Cell Description	V_{OC} (V)	J_{SC} (mA/cm^2)	FF (%)	PCE (%)
Solar cell w/o GO ARC	0.431	13.27	52.86	3.02
Solar cell with GO ARC (thickness = 69 nm)	0.502	18.62	56.72	5.31

To increase the junction built-in voltage and the open-circuit voltage of the solar cell, one suitable approach is the oxidation of the n-Si surface before the graphene transfer. Figure 6a shows the PCE dependence on the thickness δ of the SiO$_2$ layer grown by the Si surface oxidation. The PCE monotonically increases as δ increases, but a δ value exceeding 1.5 nm should be avoided because it could prevent the carrier transport through the oxide layer. Therefore, we have chosen δ = 1.5 nm. Figure 6b shows the V_{OC} vs. δ plot. Our simulation confirms that, as δ increases, V_{OC} also increases.

In particular, V_{OC} increases from 0.50 to 0.65 V by growing a SiO_2 layer with a thickness in the range from 0 nm to 1.5 nm. In addition, the SiO_2 layer, as expected, does not degrade the photo-generated current density due to its nanometer thickness and, consequently, the cell short-circuit current. The effect of the SiO_2 layer with a thickness of 1.5 nm on the fill factor is an increase in this performance parameter from 56.7% to 64.0%.

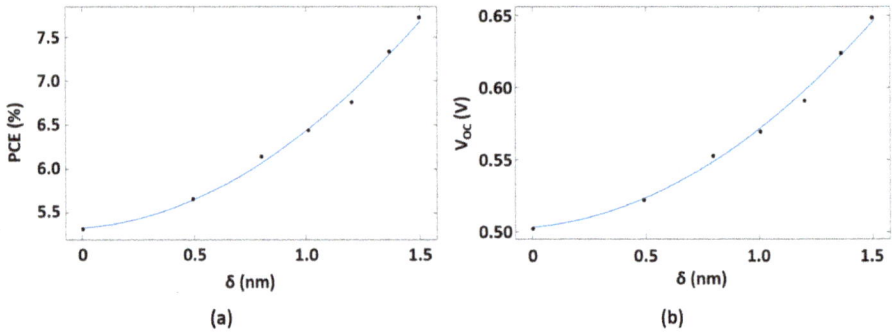

Figure 6. (**a**) PCE dependence on the thickness of the silicon oxide layer. (**b**) Open-circuit voltage vs. δ. Number of atomic layers forming the graphene sheet = 3, solar cell with GO ARC, $H = 10$ μm, and $N_D = 10^{15}$ cm^{-3}.

Several experimental results [26] have confirmed that silicon/graphene solar cells with a thickness of ≤ 50 μm are very flexible and do not alter their performance after tens of bending cycles. Thus, we investigated the possibility of increasing the thickness H of the n-Si layer to improve the cell PCE. Figure 7a shows the PCE dependence on H. We can observe that the H increase induces a PCE improvement up to 10.04% for $H = 50$ μm, mainly due to the increase of the EQE. This physical interpretation is confirmed by the J_{SC} dependence on H (see Figure 7b). The short-circuit current density, which is approximately equal to the photo-generated current density, increases as H also increases. In particular, J_{SC} varies from 18.6 mA/cm^2 to 25.7 mA/cm^2 when H increases from 10 μm to 50 μm.

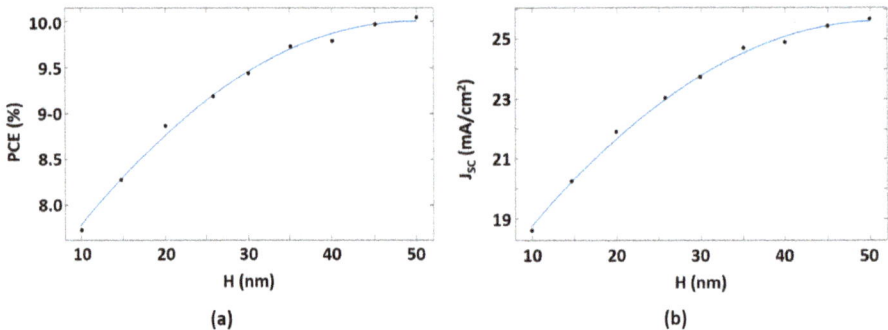

Figure 7. (**a**) PCE dependence on H. (**b**) Short-circuit current density vs. H. Number of atomic layer forming the graphene sheet = 3, solar cell with GO ARC, $\delta = 1.5$ nm, and $N_D = 10^{15}$ cm^{-3}.

The features of the optimized solar cell and its performance are summarized in Table 5.

The envisaged fabrication process of the optimized solar cell includes the patterning of the properly thinned oxidized n-Si substrate by photolithography and wet etching of SiO_2, the deposition of the metal contacts by sputtering, the immersion of the sample in ultrapure water with dissolved oxygen to grow the ultrathin SiO$_2$ layer, the graphene transfer on that layer, and the deposition of the GO layer by spin coating.

The achieved PCE improves by about 1.6% compared with the state-of-the-art flexible silicon/graphene Schottky junction solar cells, which is 8.42% [26]. To our knowledge, PCE values >10% are achievable in flexible photovoltaic devices on GaAs [41], crystalline silicon (c-Si) [42], copper indium gallium (di)selenide (CIGS) [43], and perovskite solar cells [44]. GaAs, CIGS, and c-Si cells are very stable, but their fabrication is complex, expensive, or both, while, as already mentioned, the environmental stability of perovskite solar cells is usually lower with respect to competing technologies. The values assumed for the thickness of the graphene sheet and of the SiO$_2$ layer have been already demonstrated as technologically feasible [23]. The designed flexible solar cell has a fabrication process simpler and cheaper than GaAs, CIGS, and c-Si cells; based on several experimental data [25,26], we expect that it is more stable than perovskite cells. In fact, experiments reported in [25] show that the performance of graphene/silicon Schottky junction flexible solar cells is degraded by only 0.4% after they are exposed to air for 20 days. In addition, cells based on the same technology are quite insensitive to tens (up to 50) of bending cycles (PCE degradation <0.1%) [26].

Table 5. Features and performance of the optimized solar cell.

Parameter	Symbol	Value
Doping of the n-Si layer	N_D	10^{15} cm^{-3}
Number of atomic layer forming the Graphene sheet	-	3
Thickness of the GO ARC	-	69 nm
Thickness of the SiO$_2$ layer	δ	1.5 nm
Thickness of the n-Si layer	H	50 μm
Short-circuit current density	J_{SC}	25.71 mA/cm^2
Open-circuit voltage	V_{OC}	0.66 V
Fill factor	FF	59%
Power conversion efficiency	PCE	10.04%

4. Conclusions

A new flexible graphene-on-silicon Schottky junction solar cell with a PCE of about 10% has been modeled and designed. This efficiency value improves the state of the art by more than 1.6%. The device has been optimized using a complete mathematical model we have developed. To enhance the solar cell performance, a graphene oxide layer has been used to p-dope the graphene sheet and to reduce the light reflection at the silicon/graphene interface. The open-circuit voltage and consequently the PCE of the solar cell have been increased by an ultrathin silicon oxide layer between the graphene sheet and the n-Si layer, whose thickness has been optimized through a compromise between two opposite needs, i.e., the cell efficiency and its flexibility. The designed cell can be easily manufactured by standard technological processes, and its environmental stability is better with respect to the low-cost competing technologies, such as those based on perovskites. These features of the proposed device enable its potential application in the field of IoT and wearable technology.

Author Contributions: Francesco Dell'Olio supervised the design and wrote the article; Michele Palmitessa performed the mathematical development of the model and designed the device; Caterina Ciminelli conceived the study, supervised the design, and reviewed the paper with particular reference to the physical aspects of the device.

Conflicts of Interest: The authors declare no conflict of interest.

References

1. Warner, J.H.; Schaffel, F.; Rummeli, M.; Bachmatiuk, A. *Graphene: Fundamentals and Emergent Applications*; Elsevier: Waltham, MA, USA, 2013.
2. Choi, W.; Lee, J. *Graphene: Synthesis and Applications*; CRC Press: Boca Raton, FL, USA, 2016.

3. Novoselov, K.S.; Fal'ko, V.I.; Colombo, L.; Gellert, P.R.; Schwab, M.G.; Kim, K. A roadmap for graphene. *Nature* **2012**, *490*, 192–200. [CrossRef] [PubMed]

4. Bonaccorso, F.; Sun, Z.; Hasan, T.; Ferrari, A.C. Graphene photonics and optoelectronics. *Nat. Photonics* **2010**, *4*, 611–622. [CrossRef]

5. Liu, M.; Yin, X.; Ulin-Avila, E.; Geng, B.; Zentgraf, T.; Ju, L. A graphene-based broadband optical modulator. *Nature* **2011**, *474*, 64–67. [CrossRef] [PubMed]

6. Xia, F.; Mueller, T.; Lin, Y.; Valdes-Garcia, A.; Avouris, P. Ultrafast graphene photodetector. *Nat. Nanotechnol.* **2009**, *4*, 839–843. [CrossRef] [PubMed]

7. Bao, Q.; Zhang, H.; Wang, Y.; Ni, Z.; Yan, Y.; Shen, Z.X.; Loh, K.P.; Tang, D.Y. Atomic-Layer Graphene as a Saturable Absorber for Ultrafast Pulsed Lasers. *Adv. Funct. Mater.* **2009**, *19*, 3077–3083. [CrossRef]

8. Yahiaoui, R.; Guillet, J.P.; de Miollis, F.; Mounaix, P. Ultra-flexible multiband terahertz metamaterial absorber for conformal geometry applications. *Opt. Lett.* **2013**, *38*, 4988–4990. [CrossRef] [PubMed]

9. Yahiaoui, R.; Hanai, K.; Takano, K.; Nishida, T.; Miyamaru, F.; Nakajima, M.; Hangyo, M. Trapping waves with terahertz metamaterial absorber based on isotropic Mie resonators. *Opt. Lett.* **2013**, *40*, 3197–3200. [CrossRef] [PubMed]

10. Bao, Q.; Zhang, H.; Wang, B.; Ni, Z.; Lim, C.H.Y.X.; Wang, Y.; Tang, D.Y.; Loh, K.P. Broadband graphene polarizer. *Nat. Photonol.* **2011**, *5*, 411–415. [CrossRef]

11. Conteduca, D.; Dell'Olio, F.; Ciminelli, C.; Armenise, M.N. Resonant Graphene-Based Tunable Optical Delay Line. *IEEE Photonics J.* **2015**, *7*, 7802409. [CrossRef]

12. Tatoli, T.; Conteduca, D.; Dell'Olio, F.; Ciminelli, C.; Armenise, M.N. Graphene-based fine-tunable optical delay line for optical beamforming in phased-array antennas. *Appl. Opt.* **2016**, *55*, 4342–4349. [CrossRef] [PubMed]

13. Capmany, J.; Domenech, D.; Muñoz, P. Silicon graphene waveguide tunable broadband microwave photonics phase shifter. *Opt. Express* **2014**, *22*, 8094–8100. [CrossRef] [PubMed]

14. Yin, Z.; Zhu, J.; He, Q.; Cao, X.; Tan, C.; Chen, H.; Yan, Q.; Zhang, H. Graphene-Based Materials for Solar Cell Applications. *Adv. Energy Mater.* **2014**, *4*, 1300574. [CrossRef]

15. Loh, K.P.; Tong, S.W.; Wu, J. Graphene and Graphene-like Molecules: Prospects in Solar Cells. *J. Am. Chem. Soc.* **2016**, *138*, 1095–1102. [CrossRef] [PubMed]

16. You, P.; Liu, Z.; Tai, Q.; Liu, S.; Yan, F. Efficient Semitransparent Perovskite Solar Cells with Graphene Electrodes. *Adv. Mater.* **2015**, *27*, 3632–3638. [CrossRef] [PubMed]

17. Park, H.; Chang, S.; Zhou, X.; Kong, J.; Palacios, T.; Gradečak, S. Flexible Graphene Electrode-Based Organic Photovoltaics with Record-High Efficiency. *Nano Lett.* **2014**, *14*, 5148–5154. [CrossRef] [PubMed]

18. Tong, S.W.; Wang, Y.; Zheng, Y.; Ng, M.; Loh, K.P. Graphene Intermediate Layer in Tandem Organic Photovoltaic Cells. *Adv. Funct. Mater.* **2011**, *21*, 4430–4435. [CrossRef]

19. Ameen, S.; Shaheer Akhtar, M.; Seo, H.-K.; Nazeeruddin, M.K.; Shin, H.S. An Insight into Atmospheric Plasma Jet Modified ZnO Quantum Dots Thin Film for Flexible Perovskite Solar Cell: Optoelectronic Transient and Charge Trapping Studies. *J. Phys. Chem. C* **2015**, *119*, 10379–10390. [CrossRef]

20. Wang, J.T.; Ball, J.M.; Barea, E.M.; Abate, A.; Alexander-Webber, J.A.; Huang, J.; Saliba, M.; Mora-Sero, I.; Bisquert, J.; Snaith, H.J.; et al. Low-Temperature Processed Electron Collection Layers of Graphene/TiO$_2$ Nanocomposites in Thin Film Perovskite Solar Cells. *Nano Lett.* **2014**, *14*, 724–730. [CrossRef] [PubMed]

21. Gao, P.; Ding, K.; Wang, Y.; Ruan, K.; Diao, S.; Zhang, Q.; Sun, B.; Jie, J. Crystalline Si/Graphene Quantum Dots Heterojunction Solar Cells. *J. Phys. Chem. C* **2014**, *118*, 5164–5171. [CrossRef]

22. Li, X.; Zhu, H.; Wang, K.; Cao, A.; Wei, J.; Li, C.; Jia, Y.; Li, Z.; Li, X.; Wu, D. Graphene-On-Silicon Schottky Junction Solar Cells. *Adv. Mater.* **2010**, *22*, 2743–2748. [CrossRef] [PubMed]

23. Song, Y.; Li, Y.; Mackin, C.; Zhang, X.; Fang, W.; Palacios, T.; Zhu, H.; Kong, J. Role of Interfacial Oxide in High-Efficiency Graphene—Silicon Schottky Barrier Solar Cells. *Nano Lett.* **2015**, *15*, 2104–2110. [CrossRef] [PubMed]

24. Li, X.; Chen, W.; Zhang, S.; Wu, Z.; Wang, P.; Xu, Z.; Chen, H.; Yin, W.; Zhong, H.; Lin, S. 18.5% efficient graphene/GaAs van der Waals heterostructure solar cell. *Nano Energy* **2015**, *16*, 310–319. [CrossRef]

25. Jiao, T.; Wei, D.; Liu, J.; Sun, W.; Jia, S.; Zhang, W.; Feng, Y.; Shia, H.; Dua, C. Flexible solar cells based on graphene-ultrathin silicon Schottky junction. *RSC Adv.* **2015**, *5*, 73202–73206. [CrossRef]

26. Ruan, K.; Ding, K.; Wang, Y.; Diao, S.; Shao, Z.; Zhang, X.; Jie, J. Flexible graphene/silicon heterojunction solar cells. *J. Mater. Chem. A* **2015**, *3*, 14370–14377. [CrossRef]

27. Pagliaro, M.; Palmisano, G.; Ciriminna, R. *Flexible Solar Cells*; Wiley VCH: Weinheim, Germany, 2008.

28. Lin, Q.; Huang, H.; Jing, Y.; Fu, H.; Chang, P.; Li, D.; Yao, Y.; Fan, Z. Flexible photovoltaic technologies. *J. Mater. Chem. C* **2014**, *2*, 1233–1247. [CrossRef]

29. Yin, X.; Chen, P.; Que, M.; Xing, Y.; Que, W.; Niu, C.; Shao, J. Highly Efficient Flexible Perovskite Solar Cells Using Solution-Derived NiO_x Hole Contacts. *ACS Nano* **2016**, *10*, 3630–3636. [CrossRef] [PubMed]

30. Berhe, T.A.; Su, W.N.; Chen, C.H.; Pan, C.-J.; Cheng, J.-H.; Chen, H.-M.; Tsai, M.-C.; Chen, L.Y.; Dubale, A.A.; Hwang, B.J. Organometal halide perovskite solar cells: Degradation and stability. *Energy Environ. Sci.* **2016**, *9*, 323–356. [CrossRef]

31. Yavuz, S.; Kuru, C.; Choi, D.; Kargar, A.; Jina, S.; Bandaru, P.R. Graphene oxide as a p-dopant and an anti-reflection coating layer, in graphene/silicon solar cells. *Nanoscale* **2016**, *8*, 6473–6478. [CrossRef] [PubMed]

32. Morita, M.; Ohmi, T.; Hasegawa, E.; Kawakami, M.; Ohwada, M. Growth of native oxide on a silicon surface. *J. Appl. Phys.* **1990**, *68*, 1272–1281. [CrossRef]

33. Hovel, H.J. Solar Cells. *Semicond. Semimet.* **1975**, *11*, 112–148.

34. Sze, S.M. *Physics of Semiconductor Devices*; Wiley: New York, NY, USA, 1969.

35. Green, M.A.; Keevers, M.J. Optical properties of intrinsic silicon at 300 K. *Prog. Photovolt. Res. Appl.* **1995**, *3*, 189–192. [CrossRef]

36. Fossum, J.G. Computer-aided numerical analysis of silicon solar cells. *Solid State Electron.* **1976**, *19*, 269–277. [CrossRef]

37. Arora, N.D.; Hauser, J.R.; Roulston, D.J. Electron and hole mobilities in silicon as a function of concentration and temperature. *IEEE Trans. Electron. Devices* **1982**, *2*, 292–295. [CrossRef]

38. Chen, M.J.; Wu, C.Y. A new method for computer-aided optimization of solar cell structures. *Solid State Electron.* **1985**, *28*, 751–761. [CrossRef]

39. Saleh, B.A.E.; Teich, M.C. *Fundamentals of Photonics*; Wiley: Hoboken, NJ, USA, 2007.

40. Goodstal, G. *Electrical Theory for Renewable Energy*; Delmar, Cengage Learning: Clifton Park, NY, USA, 2013.

41. Takamoto, T.; Kaneiwa, M.; Imaizumi, M.; Yamaguchi, M. InGaP/GaAs-based Multijunction Solar Cells. *Prog. Photovolt. Res. Appl.* **2005**, *13*, 495–511. [CrossRef]

42. Cruz-Campa, J.L.; Okandan, M.; Resnick, P.J.; Clews, P.; Pluym, T.; Grubbs, R.K.; Gupta, V.P.; Zubia, D.; Nielson, G.N. Microsystems enabled photovoltaics: 14.9% efficient 14 µm thick crystalline silicon solar cell. *Sol. Energy Mater. Sol. Cells* **2011**, *95*, 551–558. [CrossRef]

43. Chirilă, A.; Buecheler, S.; Pianezzi, F.; Bloesch, P.; Gretener, C.; Uhl, A.R.; Fella, C.; Kranz, L.; Perrenoud, J.; Seyrling, S.; et al. Highly efficient $Cu(In,Ga)Se_2$ solar cells grown on flexible polymer films. *Nat. Mater.* **2011**, *10*, 857–861. [CrossRef] [PubMed]

44. Halim, H.; Guo, Y. Flexible organic-inorganic hybrid perovskite solar cells. *Sci. China Mater.* **2016**, *59*, 495–506. [CrossRef]

electronics

MDPI

Article

Atomic Layer Growth of InSe and Sb₂Se₃ Layered Semiconductors and Their Heterostructure

Robert Browning *, Neal Kuperman, Bill Moon and Raj Solanki *

Department of Physics, Portland State University, Portland, OR 97207, USA; nealwkuperman@gmail.com (N.K.); bill@pdx.edu (B.M.)
* Correspondence: robertb@pdx.edu (R.B.); solanki@pdx.edu (R.S.);
  Tel.: +1-503-725-3231 (R.B.); +1-503-725-3231 (R.S.)

Academic Editors: Yoke Khin Yap and Zhixian Zhou
Received: 12 February 2017; Accepted: 27 March 2017; Published: 30 March 2017

Abstract: Metal chalcogenides based on the C–M–M–C (C = chalcogen, M = metal) structure possess several attractive properties that can be utilized in both electrical and optical devices. We have shown that specular, large area films of γ-InSe and Sb₂Se₃ can be grown via atomic layer deposition (ALD) at relatively low temperatures. Optical (absorption, Raman), crystalline (X-ray diffraction), and composition (XPS) properties of these films have been measured and compared to those reported for exfoliated films and have been found to be similar. Heterostructures composed of a layer of γ-InSe (intrinsically n-type) followed by a layer of Sb₂Se₃ (intrinsically p-type) that display diode characteristics were also grown.

Keywords: atomic layer deposition; heterostructure; pn diode

1. Introduction

The unique properties of graphene have drawn attention to other 2-dimensional (2D) materials, in particular metal chalcogenides, which can behave as metals, insulators, or semiconductors [1–3]. These semiconductors are of particular interest for the fabrication of the future generation of field effect transistors (FETs) and optoelectronic devices. These layered materials can be used as single or a few atomic layer thick films, absent of dangling bonds which allows for better electrostatic control of carrier transport; hence, they are ideal for downscaling of FETs. The ability to tune the direct band gap of many of these materials, by controlling their thickness, allows further flexibility for the design and fabrication of optoelectronic devices. However, to date, most attention has been focused on 2D transition metal dichalcogenides of structure C–M–C (C = chalcogen, M = metal), where one metal atom is sandwiched between two chalcogen atoms [4–7]. There are other forms of metal chalcogenides that are equally intriguing. Among these is the C–M–M–C family, which has been drawing more attention [8,9]. These materials have been used to produce photocatalysts, photo-detectors, image sensors, and transistors with high electron mobility [10–13]. We are currently investigating 2D semiconductor layers of this family of chalcogenide material that are composed of two metal and two Se atoms, Se–M–M–Se (M = In, Sb) grown via atomic layer deposition (ALD). Our objective is to demonstrate the growth of uniform, large area films that may in the future be utilized for the fabrication of electronic and optical devices. Our objective in this communication is to demonstrate that uniform and smooth films of InSe and Sb₂Se₃, both belonging to the C–M–M–C family of metal chalcogenides, can be grown over large areas using ALD. The properties of thin films grown using this approach, and their heterostructure are presented below.

InSe has several polytypes, however the β-phase and γ-phase are the two common forms of its crystal structure [14]. In this work, only the γ-InSe phase is considered which crystallizes into stacked hexagonal layers. The vertical stacking of this material is composed of Se–In–In–Se sheets, where each sheet is weakly bound to its neighboring sheets by the van der Waals force [14]. It is interesting to note that while bulk InSe has a direct band gap (1.2 eV), reduction of its thickness first leads to a wider band gap due to quantum confinement, however further reduction of the thickness to a few layers results in a strong decrease of the photoluminescence. One of the reasons for this attenuation has been attributed to direct-to-indirect bandgap crossover, while another explanation has been attributed to the enhancement of non-radiative recombination processes in the thin flakes [15,16]. To date, InSe films have been produced mostly via chemical vapor deposition (CVD), sputtering, exfoliation, and electrodeposition [15,17–20]. In these cases, the films are either small, on the order of a few tens of micrometers per side (exfoliation) or are composed of flakes or platelets.

Unlike InSe, Sb_2Se_3 is a narrow band gap, layered metal chalcogenide (Se–Sb–Sb–Se) that has a single phase [21]. It is a direct band gap semiconductor with a layered structure and an orthorhombic structure [22]. The films consist of staggered, parallel layers of 1D $(Sb_4Se_6)_n$ ribbons that are composed of strong Sb–Se bonds along the [001] direction. In the [100] and [010] directions, the ribbons are held in the stack by the weak van der Waals forces [23]. It has received attention as a thermoelectric material, and recently as a light sensitizer in photovoltaic devices because of its narrow band gap of about 1.1 eV–1.3 eV, which approaches the ideal Shockley–Queisser value [24–26]. These films have been produced via a number of methods, including thermal evaporation, chemical bath deposition, spray pyrolysis, pulsed laser deposition and electrochemical deposition [25,27–30].

2. Materials and Methods

InSe and Sb_2Se_3 films were grown in a Microchemistry F-120 ALD reactor (Microchemistry, Helsinki, Finland) that can handle two 50 mm $\times$ 50 mm substrates per run. The substrates for these films consisted of either p-or n-type Si wafers coated with a 320 nm thick film of thermal silicon oxide. Besides Si, plain glass slides and fluorinated tin oxide (FTO) coated glass plates were also used as substrates. The precursors for InSe growth were $InCl_3$ and H_2Se (8%, balance Ar). The In source temperature was set at 110 °C and the carrier gas was nitrogen. The pulse sequence per cycle was as follows; $InCl_3$ pulse width of 1 s; N_2 purge of 1.0 s; H_2Se pulse width of 1.5 s; followed by 1.0 s N_2 purge. Uniform film growth occurred over a temperature range of 310 °C to 380 °C. The films reported here were grown at 350 °C, where the growth rate was 0.05 nm per cycle. The pulse sequence for Sb_2Se_3 films was $SbCl_3$ (1 s), N_2 (1.5 s), H_2Se (1.5 s), and N_2 (1.5 s). Uniform films were produced over a temperature range of 270 °C to 320 °C. From this range, 300 °C was selected to grow the films, where the growth rate was 0.22 nm/cycle. Optimization of growth conditions was done mostly based on Raman analysis, which offered a quick turn-around. Besides Raman, the crystalline structure of these films was examined with X-ray diffraction (XRD) (Rigaku Corp., Tokyo, Japan), their composition with X-ray photoelectron spectroscopy (XPS) (Northrop-Grumman, Falls Church, VA, USA) and their band gaps with optical absorption. The surfaces of these films were examined with a scanning electron microscope (SEM) (FEI Corp, Hillsboro, OR, USA) and found to be featureless. Finally, $InSe/Sb_2Se_3$ heterostructures were grown and their current-voltage (I–V) profiles were examined.

3. Results

The surface morphology of the films was first checked with a Nomarski microscope and then with an SEM. When the growth process was optimized, the surfaces of both InSe and Sb_2Se_3 were smooth and featureless, as shown in Figure 1a,b. This is a major difference between films grown via ALD and other methods. Although we expect there to be grain boundaries, we were unable to detect them with the SEM. Also, no grain boundaries were visible with the electron backscatter diffraction (EBSD) method.

Figure 1. Scanning electron microscope (SEM) view of the surface of (a) InSe and (b) Sb$_2$Se$_3$.

The crystalline properties of the films were characterized using grazing-angle (0.5°) XRD. The XRD spectrum of a 17 nm thick InSe film is shown in Figure 2a. The spectrum is dominated by the [006] peak, followed by minor peaks, which are characteristic of the hexagonal structure of γ-InSe (JCPDS 40-1407). Also, the deposition of an orange film on the walls of the reactor is indicative of this phase of the InSe [31].

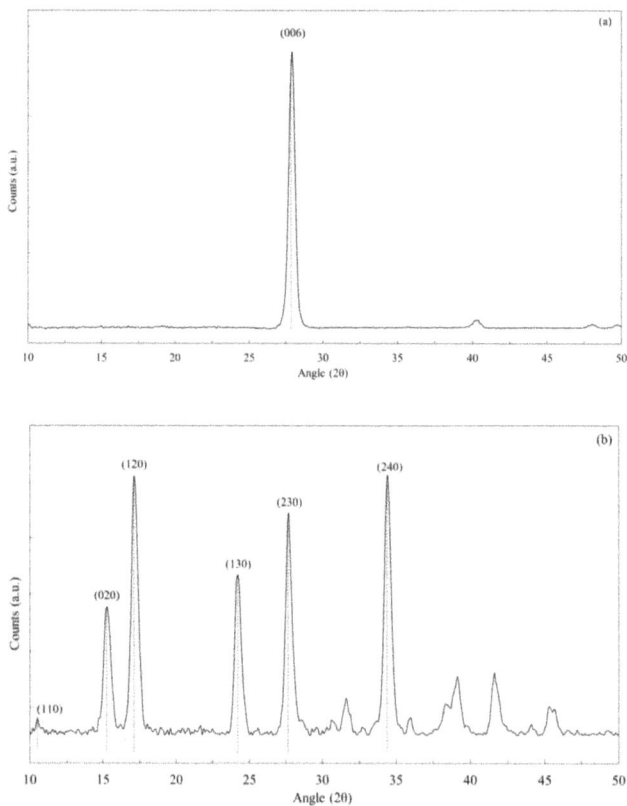

Figure 2. X-ray diffraction (XRD) spectrum of (a) InSe (b) Sb$_2$Se$_3$.

The XRD spectrum of our antimonselite (Sb$_2$Se$_3$) film is shown in Figure 2b, where all of the diffraction peaks agree well with its orthorhombic crystal structure (JCPDS 15-0861). Major peaks are indexed to the diffraction planes. The crystal lattice parameters were calculated as a = 1.159 nm, b = 1.175 nm, and c = 0.3949 nm, that are consistent with the values reported in literature [32]. No peaks of any other phases were detected, which indicates that these films are of single phase and high purity.

Optical characterization involved Raman spectroscopy and absorption spectra of these films. The Raman spectrum of a 17 nm thick InSe film is shown in Figure 3a, where the excitation wavelength was 532 nm. It shows the signature peaks corresponding to both the in-plane (E) and out-of-plane (A) modes that are consistent with those reported for flakes of γ-InSe [15,18,33,34]. Comparing the InSe data in Figure 3a with published results for the bulk and thin films, we can identify the peaks at about 117 cm^{-1}, 175 cm^{-1}, and 205 cm^{-1} with A$'_1$, E$'_{2g}$, and A$'_1$(LO) Raman modes, respectively. The peak at 205 cm^{-1} is strongest in films that are at about 6 nm to 9 nm thick layers and then attenuates with increase in thickness. The peak at around 155 cm^{-1} is distinctive for the γ phase of InSe and is related to the zone center mode of the crystal [35,36].

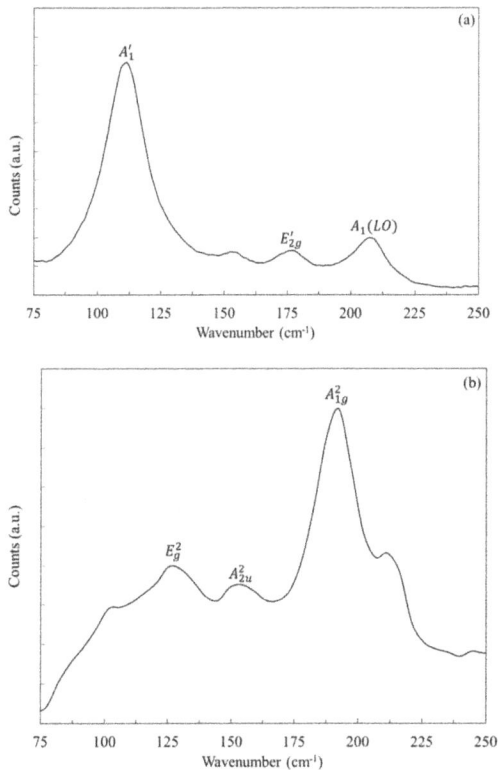

Figure 3. Raman spectrograph of (a) γ-InSe (b) Sb$_2$Se$_3$.

The Raman spectrum of the 15 nm thick Sb$_2$Se$_3$ thin film is shown in Figure 3b. It agrees well with previously reported data, where the main peak at 189 cm^{-1} (A$^2_{1g}$) is the characteristic Sb–Se stretching mode of the SbSe$_{3/2}$ pyramids. The two other peaks, at about 151 cm^{-1} (A$^2_{2u}$) and at about 125 cm^{-1} (E$_{2g}$) are associated with the Sb–Sb bonds and Se–Se bonds, respectively [37–39].

Band gaps of these films were determined from their respective absorption spectra obtained from the films deposited on glass substrates. The absorbance of these films was recorded over a spectral range of 300 to 1100 nm. Plots of the wavelength (in terms of energy E in eV) versus the absorbance InSe and Sb_2Se_3 films are plotted in Figure 4a,b, respectively. The intercept on the x-axis of the extrapolation of the linear portion of the slope corresponds to the band gap of the material. For the 17 nm thick InSe film, the intercept occurs at about 2.35 eV. This value is in good agreement with the band gap of γ-InSe, which is defined as the transition between the p_x and p_y orbital to the bottom of the conduction band that corresponds to an energy gap of 2.4 eV [15,18]. The band gap of γ-InSe is higher than the other phases, therefore this parameter can be used to discriminate it from the other phases [35]. From Figure 4b, a band gap value of 1.3 eV was determined for a 15 nm thick Sb_2Se_3 film. This value is within the range of 1.1 to 1.3 eV previously reported [21,40].

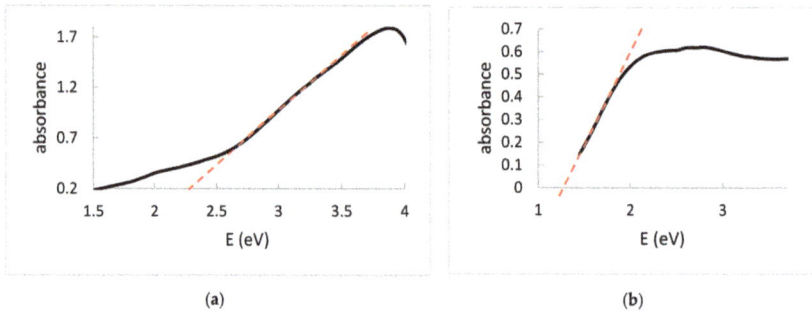

(a)

(b)

Figure 4. Absorbance spectrum of (**a**) InSe (**b**) Sb_2Se_3.

The chemical composition of both of these films was examined by X-ray photoelectron spectroscopy (XPS). The signature binding energies of In $3d_{5/2}$ and In $3d_{3/2}$, shown in Figure 5a, correspond to 452.47 eV and 444.88 eV, respectively. The Se 3d profile is deconvoluted to show the two, closely spaced binding profiles of the Se $d_{5/2}$ and $d_{3/2}$ of binding energies of 53.88 eV and 54.88 eV (Figure 5b). These values are in agreement with pure In_2Se_3 films [41]. The dip between the two peaks is an indication of oxidation of the films, which is not surprising since the samples were exposed to air for a short period of time before XPS analysis.

The XPS binding energy of Sb_2Se_3 is shown in Figure 6. The binding energies of Sb $3d_{5/2}$ and $3d_{3/2}$ are 529.83 eV and 539.03 eV as shown in Figure 6a. Detailed spectral deconvolution of the Se 3d high resolution XPS spectrum revealed that the binding energy of Se $3d_{5/2}$ and $3d_{3/2}$ are 54.43 eV and 55.23 eV, respectively, which is in good agreement with the expected binding energy of Sb_2Se_3 [25]. These binding energies were corrected by referencing the C 1s peak to 284.70 eV.

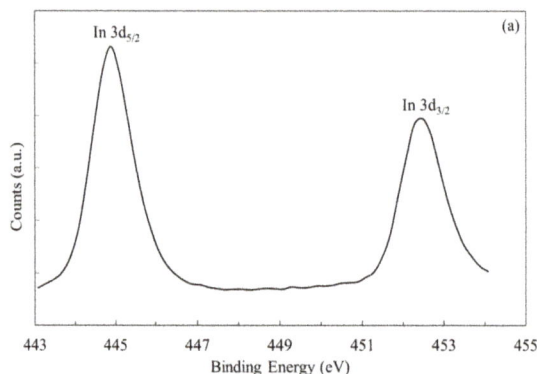

Figure 5. *Cont.*

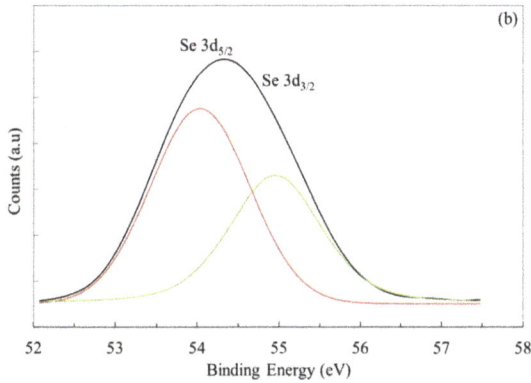

Figure 5. X-ray photoelectron spectroscopy (XPS) profile of (**a**) In (**b**) Se.

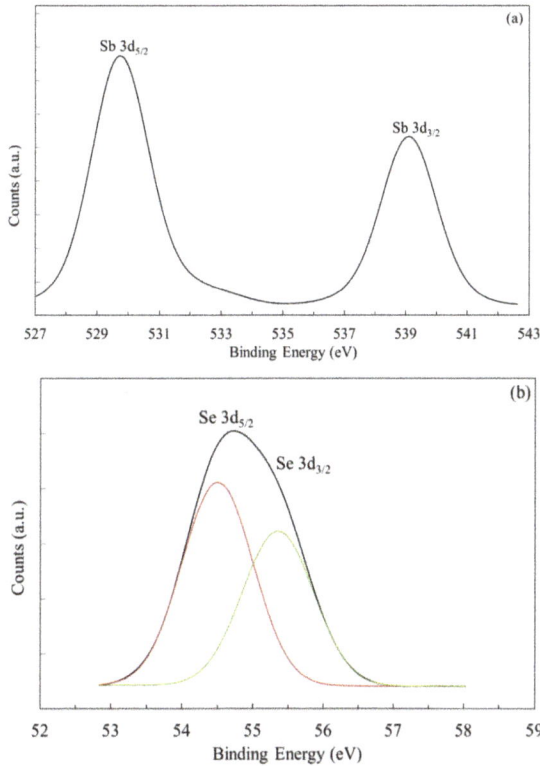

Figure 6. XPS profile of (**a**) Sb (**b**) Se.

4. Discussion

In an ALD process, the choice of the precursors and substrates is important in order to produce a saturated chemisorbed layer of the first species during the initial phase of the growth [42]. Moreover, the deposition parameters such as the pulse widths and temperature must be optimized to produce this condition. Then, the growth will proceed layer by layer, which is ideal for growth of a

2D film. We have found that organometallic or chloride-based metal precursors (with sufficient vapor pressure) work well on SiO_2 covered substrates, as we have previously demonstrated with other metal chalcogenides [5,43,44].

The 2D materials, as described above, are composed of individual layers that are held together by the weak van der Waals force. Within each layer, the atoms are strongly bonded to each other in 2D via either valence or ionic bonds, and in the absence of dangling bonds. This allows stacking of two different 2D materials with different physical properties, crystalline parameters and crystal symmetry to grow heterojunctions and superlattices held together by the van der Waals force. This approach to grow 2D heterojunctions has been referred to as van der Waals epitaxy [45]. We have grown heterostructures composed of 5 nm–10 nm thick InSe and 5 nm–7 nm thick Sb_2Se_3 layers by sequentially growing each layer on Si/SiO_2 substrates and FTO-coated glass substrates. While detailed electrical characterization of these films is underway, Hall effect was used to determine the intrinsic doping properties of each of these two films. The InSe films were found to be intrinsically n-type and Sb_2Se_3 were intrinsically p-type. On the FTO substrates, the first film deposited was InSe, followed by Sb_2Se_3, while keeping one part of the FTO surface covered to avoid any deposition. Small Au contacts were sputtered on top of the Sb_2Se_3 layer and Al on the FTO portion. The current–voltage (I–V) profiles of these samples were then examined.

A typical example of one of these samples is shown in Figure 7, which shows the characteristic profile of a pn diode. The leakage current with a reverse bias is shown (in log scale) in the inset. A maximum leakage current of 10^{-7} A was measured at -1 V bias. Since each precursor was introduced separately into the reactor for a surface reaction for a layer-by-layer growth, and we were working at a relatively low temperature, we do not expect significant intermixing at the interface. This heterojunction was grown at 310 °C, which overlaps the optimum range of both the materials. A one minute purge was introduced before starting the Sb_2Se_3 layer to ensure a sharp interface. In our previous work on SnS/WS_2 heterojunctions that were grown at a slightly higher temperature, there were no obvious signs of intermixing at the interface [5].

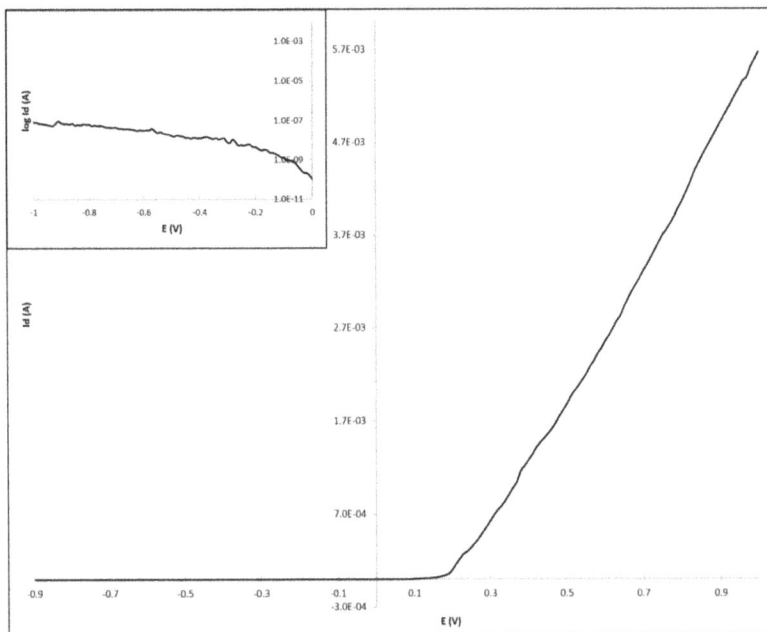

Figure 7. Current–voltage (I–V) profile of the InSe/Sb_2Se_3 pn junction.

5. Conclusions

In summary, ALD has been utilized to grow smooth and continuous layers of γ-InSe and Sb$_2$Se$_3$ layered metal chalcogenide on SiO$_2$-coated Si and glass substrates. Crystalline and optical properties, as well as the composition of these films are comparable to those produced by other means, especially exfoliation. The InSe and Sb$_2$Se$_3$ films show intrinsic n-type and p-type behaviors, respectively. A combination of these two films shows pn diode characteristics. These results show that ALD offers a viable path for producing large area films of metal chalcogenides for future industrial scale applications.

Author Contributions: Robert Browning, Neal Kuperman, Bill Moon and Raj Solanki all contributed equally.

Conflicts of Interest: The authors declare no conflict of interest.

References

1. Chhowalla, M.; Shin, H.S.; Eda, G.; Li, L.J.; Loh, K.P.; Zhang, H. The chemistry of two-dimensional layered transition metal dichalcogenide nanosheets. *Nat. Chem.* **2013**, *5*, 263–275. [CrossRef] [PubMed]
2. Xu, M.; Liang, T.; Shi, M.; Chen, H. Graphene-like two-dimensional materials. *Chem. Rev.* **2013**, *113*, 3766–3798. [CrossRef] [PubMed]
3. Geim, A.K.; Grigorieva, I.V. Van der waals heterostructures. *Nature* **2013**, *499*, 419–425. [CrossRef] [PubMed]
4. Wang, Q.H.; Kalantar-Zadeh, K.; Kis, A.; Coleman, J.N.; Strano, M.S. Electronics and optoelectronics of two-dimensional transition metal dichalcogenides. *Nat. Nanotechnol.* **2012**, *7*, 699–712. [CrossRef] [PubMed]
5. Browning, R.; Plachinda, P.; Padigi, P.; Solanki, R.; Rouvimov, S. Growth of multiple ws2/sns layered semiconductor heterojunctions. *Nanoscale* **2016**, *8*, 2143–2148. [CrossRef] [PubMed]
6. Das, S.; Chen, H.Y.; Penumatcha, A.V.; Appenzeller, J. High performance multilayer mos2 transistors with scandium contacts. *Nano Lett.* **2013**, *13*, 100–105. [CrossRef] [PubMed]
7. Radisavljevic, B.; Radenovic, A.; Brivio, J.; Giacometti, V.; Kis, A. Single-layer mos2 transistors. *Nat. Nanotechnol.* **2011**, *6*, 147–150. [CrossRef] [PubMed]
8. Li, X.; Lin, M.W.; Puretzky, A.A.; Idrobo, J.C.; Ma, C.; Chi, M.; Yoon, M.; Rouleau, C.M.; Kravchenko, I.I.; Geohegan, D.B.; et al. Controlled vapor phase growth of single crystalline, two-dimensional gase crystals with high photoresponse. *Sci. Rep.* **2014**, *4*, 5497. [PubMed]
9. Luxa, J.; Wang, Y.; Sofer, Z.; Pumera, M. Layered post-transition-metal dichalcogenides (x-m-m-x) and their properties. *Chemistry* **2016**, *22*, 18810–18816. [CrossRef] [PubMed]
10. Feng, W.; Zheng, W.; Cao, W.; Hu, P. Back gated multilayer inse transistors with enhanced carrier mobilities via the suppression of carrier scattering from a dielectric interface. *Adv. Mater.* **2014**, *26*, 6587–6593. [CrossRef] [PubMed]
11. Lei, S.; Wen, F.; Li, B.; Wang, Q.; Huang, Y.; Gong, Y.; He, Y.; Dong, P.; Bellah, J.; George, A.; et al. Optoelectronic memory using two-dimensional materials. *Nano Lett.* **2015**, *15*, 259–265. [CrossRef] [PubMed]
12. Harvey, A.; Backes, C.; Gholamvand, Z.; Hanlon, D.; McAteer, D.; Nerl, H.C.; McGuire, E.; Seral-Ascaso, A.; Ramasse, Q.M.; McEvoy, N.; et al. Preparation of gallium sulfide nanosheets by liquid exfoliation and their application as hydrogen evolution catalysts. *Chem. Mater.* **2015**, *27*, 3483–3493. [CrossRef]
13. Hu, P.; Wen, Z.; Wang, L.; Tan, P.; Xiao, K. Synthesis of few-layer gase nanosheets for high performance photodetectors. *ACS Nano* **2012**, *6*, 5988–5994. [CrossRef] [PubMed]
14. Amory, C.; Bernède, J.C.; Marsillac, S. Study of a growth instability of γ-in2se3. *J. Appl. Phys.* **2003**, *94*, 6945–6948. [CrossRef]
15. Sánchez-Royo, J.F.; Muñoz-Matutano, G.; Brotons-Gisbert, M.; Martínez-Pastor, J.P.; Segura, A.; Cantarero, A.; Mata, R.; Canet-Ferrer, J.; Tobias, G.; Canadell, E.; et al. Electronic structure, optical properties, and lattice dynamics in atomically thin indium selenide flakes. *Nano Res.* **2014**, *7*, 1556–1568. [CrossRef]
16. Mudd, G.W.; Svatek, S.A.; Ren, T.; Patane, A.; Makarovsky, O.; Eaves, L.; Beton, P.H.; Kovalyuk, Z.D.; Lashkarev, G.V.; Kudrynskyi, Z.R.; et al. Tuning the bandgap of exfoliated inse nanosheets by quantum confinement. *Adv. Mater.* **2013**, *25*, 5714–5718. [CrossRef] [PubMed]
17. Shigeru, S.; Tesuo, I. Crystalline inse films prepared by rf-sputtering technique. *Jpn. J. Appl. Phys.* **1991**, *30*, L2127.

18. Lei, S.; Ge, L.; Najmaei, S.; George, A.; Kappera, R.; Lou, J.; Chhowalla, M.; Yamaguchi, H.; Gupta, G.; Vajtai, R.; et al. Evolution of the electronic band structure and efficient photo-detection in atomic layers of inse. *ACS Nano* **2014**, *8*, 1263–1272. [CrossRef] [PubMed]

19. Stoll, S.L.; Barron, A.R. Metal− organic chemical vapor deposition of indium selenide thin films. *Chem. Mater.* **1998**, *10*, 650–657. [CrossRef]

20. Ho, C.H.; Lin, C.H.; Wang, Y.P.; Chen, Y.C.; Chen, S.H.; Huang, Y.S. Surface oxide effect on optical sensing and photoelectric conversion of alpha-in2se3 hexagonal microplates. *ACS Appl. Mater. Interfaces* **2013**, *5*, 2269–2277. [CrossRef] [PubMed]

21. Ghosh, G. The sb-se (antimony-selenium) system. *J. Phase Equilibria* **1993**, *14*, 753–763. [CrossRef]

22. Chang, H.-W.; Sarkar, B.; Liu, C.W. Synthesis of sb2se3 nanowires via a solvothermal route from the single source precursor sb[se2p(oipr)2]3. *Cryst. Growth Des.* **2007**, *7*, 2691–2695. [CrossRef]

23. Zhou, Y.; Wang, L.; Chen, S.; Qin, S.; Liu, X.; Chen, J.; Xue, D.-J.; Luo, M.; Cao, Y.; Cheng, Y.; et al. Thin-film sb2se3 photovoltaics with oriented one-dimensional ribbons and benign grain boundaries. *Nat. Photonics* **2015**, *9*, 409–415. [CrossRef]

24. Kutasov, V.A. Shifting the maximum figure-of-merit of (bi, sb)2(te, se)3 thermoelectrics to lower temperatures. In *Thermoelectrics Handbook*; CRC Press: Boca Raton, FL, USA, 2005; pp. 37-18–37-31.

25. Liu, X.; Chen, J.; Luo, M.; Leng, M.; Xia, Z.; Zhou, Y.; Qin, S.; Xue, D.J.; Lv, L.; Huang, H.; et al. Thermal evaporation and characterization of sb2se3 thin film for substrate sb2se3/cds solar cells. *ACS Appl. Mater. Interfaces* **2014**, *6*, 10687–10695. [CrossRef] [PubMed]

26. Shockley, W.; Queisser, H.J. Detailed balance limit of efficiency of p-n junction solar cells. *J. Appl. Phys.* **1961**, *32*, 510–519. [CrossRef]

27. Rajpure, K.; Bhosale, C. Effect of se source on properties of spray deposited sb 2 se 3 thin films. *Mater. Chem. Phys.* **2000**, *62*, 169–174. [CrossRef]

28. Rodriguez-Lazcano, Y.; Peña, Y.; Nair, M.; Nair, P. Polycrystalline thin films of antimony selenide via chemical bath deposition and post deposition treatments. *Thin Solid Films* **2005**, *493*, 77–82. [CrossRef]

29. Fernandez, A.; Merino, M. Preparation and characterization of sb 2 se 3 thin films prepared by electrodeposition for photovoltaic applications. *Thin Solid Films* **2000**, *366*, 202–206. [CrossRef]

30. Xue, M.-Z.; Fu, Z.-W. Pulsed laser deposited sb 2 se 3 anode for lithium-ion batteries. *J. Alloys Compd.* **2008**, *458*, 351–356. [CrossRef]

31. Tabernor, J.; Christian, P.; O'Brien, P. A general route to nanodimensional powders of indium chalcogenides. *J. Mater. Chem.* **2006**, *16*, 2082. [CrossRef]

32. Tideswell, N.W.; Kruse, F.H.; McCullough, J.D. The crystal structure of antimony selenide, sb2se3. *Acta Crystallogr.* **1957**, *10*, 99–102. [CrossRef]

33. Chen, Z.; Gacem, K.; Boukhicha, M.; Biscaras, J.; Shukla, A. Anodic bonded 2d semiconductors: From synthesis to device fabrication. *Nanotechnology* **2013**, *24*, 415708. [CrossRef] [PubMed]

34. Kumazaki, K.; Imai, K. Far-infrared reflection and raman scattering spectra in γ-inse. *Phys. Status Solidi* **1988**, *149*, K183–K186. [CrossRef]

35. Marsillac, S.; Combot-Marie, A.M.; Bernède, J.C.; Conan, A. Experimental evidence of the low-temperature formation of γ-in2se3 thin films obtained by a solid-state reaction. *Thin Solid Films* **1996**, *288*, 14–20. [CrossRef]

36. Weszka, J.; Daniel, P.; Burian, A.; Burian, A.M.; Nguyen, A.T. Raman scattering in in2se3 and inse2 amorphous films. *J. Non-Cryst. Solids* **2000**, *265*, 98–104. [CrossRef]

37. Bera, A.; Pal, K.; Muthu, D.V.S.; Sen, S.; Guptasarma, P.; Waghmare, U.V.; Sood, A.K. Sharp raman anomalies and broken adiabaticity at a pressure induced transition from band to topological insulator in sb2se3. *Phys. Rev. Lett.* **2013**, *110*, 107401. [CrossRef] [PubMed]

38. Ivanova, Z.G.; Cernoskova, E.; Vassilev, V.S.; Boycheva, S.V. Thermomechanical and structural characterization of gese2–sb2se3–znse glasses. *Mater. Lett.* **2003**, *57*, 1025–1028. [CrossRef]

39. Wang, J.; Deng, Z.; Li, Y. Synthesis and characterization of sb2se3 nanorods. *Mater. Res. Bull.* **2002**, *37*, 495–502. [CrossRef]

40. Filip, M.R.; Patrick, C.E.; Giustino, F. Gw quasiparticle band structures of stibnite, antimonselite, bismuthinite, and guanajuatite. *Phys. Rev. B* **2013**, *87*, 2450–2458. [CrossRef]

41. Beck, K.M.; Wiley, W.R.; Venkatasubramanian, E.; Ohuchi, F. Vacancies ordered in screw form (vosf) and layered indium selenide thin film deposition by laser back ablation. *Appl. Surf. Sci.* **2009**, *255*, 9707–9711. [CrossRef]

42. Suntola, T. Atomic layer epitaxy. *Mater. Sci. Rep.* **1989**, *4*, 261–312. [CrossRef]
43. Browning, R.; Padigi, P.; Solanki, R.; Tweet, D.J.; Schuele, P.; Evans, D. Atomic layer deposition of mos2 thin films. *Mater. Res. Express* **2015**, *2*, 035006. [CrossRef]
44. Browning, R.; Kuperman, N.; Solanki, R.; Kanzyuba, V.; Rouvimov, S. Large area growth of layered wse2 films. *Semicond. Sci. Technol.* **2016**, *31*, 095002. [CrossRef]
45. Saiki, K.; Ueno, K.; Shimada, T.; Koma, A. Application of van der waals epitaxy to highly heterogeneous systems. *J. Cryst. Growth* **1989**, *95*, 603–606. [CrossRef]

MDPI AG

St. Alban-Anlage 66

4052 Basel, Switzerland

Tel. +41 61 683 77 34

Fax +41 61 302 89 18

http://www.mdpi.com

Electronics Editorial Office

E-mail: electronics@mdpi.com

http://www.mdpi.com/journal/electronics